Crime Sc

DONALD O. SCHULTZ
Broward Community College

Crime Scene Investigation

Prentice-Hall, Inc., Englewood Cliffs, New Jersey

Library of Congress Cataloging in Publication Data

SCHULTZ, DONALD O., (date)
   Crime scene investigation.

    Bibliography: p.
    Includes index.
    1. Criminal investigation. 2. Crime laboratories. I. Title
HV8073.S345   364.12'1   76-7559
ISBN 0-13-192864-3

Prentice-Hall Series in Criminal Justice
James D. Stinchcomb, Editor

© 1977 by Prentice-Hall, Inc., Englewood Cliffs, N.J.

All rights reserved. No part of this book may be reproduced in any form or by any means without permission in writing from the publisher.

Printed in the United States of America

10 9 8 7 6 5 4 3 2 1

Prentice-Hall International, Inc., London
Prentice-Hall of Australia Pty. Limited, Sydney
Prentice-Hall of Canada, Ltd., Toronto
Prentice-Hall of India Private Limited, New Delhi
Prentice-Hall of Japan, Inc., Tokyo
Prentice-Hall of Southeast Asia Pte. Ltd. Singapore
Whitehall Books Limited, Wellington, New Zealand

Dedicated to

James F. McGowan
Lt. Col., U.S. Army (Retired),
First Director,
Criminal Justice Institute,
Broward Community College,
Fort Lauderdale, Florida

Contents

PREFACE / xi

1 CRIME SCENE PROCESSING / 1

Introduction / 1
Protecting the Crime Scene and Evidence / 1
Action at the Scene / 2
Sketches and Notes / 8

2 HANDLING AND CARE OF PHYSICAL EVIDENCE / 15

Introduction / 15
Definitions / 16
Identification of Physical Evidence / 16
Chain of Custody / 18
Disposition of Evidence / 23
Packaging and Transmitting Evidence / 24
Packing and Transmittal Chart / 26

3 TRACE EVIDENCE / 46

Introduction / 46
Toolmarks / 47
Serial Numbers / 59
Jewelers' Marks / 61
Laundry and Dry-Cleaning Marks / 61
Bloodstain Evidence / 62
Other Body Fluids / 67
Hairs and Fibers / 70
Fingernail Scrapings / 74
Soils, Rocks, and Minerals / 76
Miscellaneous Laboratory Examinations / 81

4 PHOTOGRAPHY / 86

Introduction / 86
Cameras / 86
Exposure Meters / 89
Film / 89
Filters / 92
Infrared and Ultraviolet Photography / 93
Color Photography / 96
Photographing Crime Scenes, and Other Uses of Photography / 98
Legal Photography / 102

5 CASTING AND MOLDING / 104

Introduction / 104
Definitions / 105
Materials Required / 105
Preservation and Recording of Impressions / 106
Casting / 108
Molding / 117
Forwarding to Laboratory / 118
Laboratory Examination and Analysis / 119

6 FINGERPRINTING / 120

Introduction / 120
Characteristics of Fingerprints / 120
Definitions / 121
Fingerprint Records and Forms / 121
Equipment / 122
Recording Fingerprints / 123
Problem Prints / 127
Latent Fingerprints / 129
Fingerprint Patterns / 141
Maintenance of Fingerprint Records / 146

7 FIREARMS IDENTIFICATION / 147

Introduction / 147
Definitions / 147
Responsibilities / 149
Evidence / 150
Laboratory Examination / 160

8 GLASS FRACTURES / 166

Introduction / 166
Examination of Glass Evidence / 166
The Nature and Properties of Glass / 167
Collection and Handling / 168
Evaluation of Fragments and Fractures / 170
Reconstruction of Fractured Glass / 180
Paint Spots on Glass / 182
Laboratory Examination and Analysis / 182
Forwarding to the Laboratory / 185

9 QUESTIONED DOCUMENTS AND RELATED OFFENSES / 186

Introduction / *186*
Definitions / *186*
Investigation / *187*
Offenses Involving Questioned Documents / *189*
Investigative Considerations / *190*
Miscellaneous Document Examination / *202*

10 DRUGS / 205

Introduction / *205*
Definitions / *205*
Characteristics of Drugs / *208*
Investigative Considerations / *226*

BIBLIOGRAPHY / 238

INDEX / 241

Preface

It is important that all police personnel assigned to patrol, vice, traffic accident investigation, and the investigation division have some working knowledge of basic crime lab techniques. The principal or assistant at a crime scene must possess the knowledge and skill to preserve and process the evidence available there. Quite often, the success or failure of the entire investigation will depend on proper handling of the crime scene and proper processing of the evidence.

Because of the complexity of crime and the increasing need for technical knowledge, the law enforcement community is becoming more dependent on its crime lab technicians. Only a few years ago, it was still the practice of many medium- and large-sized police agencies to rely on outside crime laboratories to process their evidence. Recently, however, because of time factors and overloaded work schedules at the major crime labs, many police agencies have had to turn to their own personnel to perform this important specialty function.

Crime Scene Investigation is a text for criminal justice students, police officers, and potential lab technicians. It is a primer in the areas of crime scene processing, handling and care of physical evidence, trace evidence, photography, casting and molding, fingerprinting, firearms identification, glass fractures, questioned documents, and drugs.

Many of the techniques found in this text are based upon the U.S. Army Field Manual. The basic research and assistance given to civilian police departments by the U.S. Army Crime Lab and its personnel is a debt which would be difficult to repay.

1

Crime Scene Processing

INTRODUCTION

The processing of a crime scene requires the application of diligent and careful methods by investigators to recognize, identify, preserve, and collect facts and items of evidentiary value that may assist in reconstructing what actually occurred. The area processed as the scene is the area that includes all direct traces of the crimes, as determined by the type of crime committed and the place where the act occurred, including the surrounding area.

In this chapter, the steps to be taken in processing the crime scene will be mentioned and described only briefly; they will be covered in more detail in subsequent chapters.

PROTECTING THE CRIME SCENE AND EVIDENCE

Successful crime scene processing depends upon the investigator's skill in recognizing and collecting facts and items of value as evidence, and upon his ability to protect, preserve, and later present them in a logical manner. This requires making careful and detailed notes and sketches; using correct procedures in taking photographs of the scene; taking written statements and transcribing verbal statements of witnesses, suspects, and victims; and carefully identifying, marking, and preserving collected physical objects of an evidentiary nature.

ACTION AT THE SCENE

Early Action

Normally when a crime has occurred, police patrols will have arrived at the scene prior to the investigator and will have taken steps to protect the scene and evacuate the seriously injured. When the investigator arrives, he should take the following measures:

1. Record the date, arrival time, and weather conditions.

2. If an injured person is on the scene, arrange for medical attention, identification, and removal. The scene is disturbed only to the extent necessary to have medical aid rendered to the injured or to have a doctor examine a deceased victim. Such alteration of the scene should be accurately recorded.

3. If the offender is at the scene, apprehend him.

4. If the scene is not fully protected, ensure its protection by using police personnel or other responsible persons to keep the curious away from it and to keep witnesses, suspects, and victims who are present from disturbing it.

It may be necessary to reroute traffic, prohibit employees from entering their place of employment, or take other action to prevent any disturbance of the scene until a complete examination is made. The body of a deceased victim should not be covered until thoroughly processed for evidence. Premature covering could result in destruction of valuable evidence.

Early action is taken to protect items of possible evidentiary value that may be destroyed by rain, snow, fire, or other causes before collection can be effected. For example, a raincoat or piece of canvas may be used to cover impressions in the earth that are exposed to rain. Wooden or pasteboard boxes may be placed over impressions in snow. Items that will melt should be shielded from the sun or other heat sources. Objects such as food and blood should be covered to protect them from contamination.

5. Determine and record the names of those at the scene who may be witnesses, and separate them. These people should be removed from the immediate area of the scene as soon as practicable.

6. Conduct preliminary questioning of witnesses, suspects, and victim(s) to determine in general the extent of the incident or crime.

7. People present within the immediate area of the scene should be only the minimum number needed to assist the investigator. It may be necessary to request others present, such as news reporters, to refrain from examining or disturbing objects or aspects of the scene.

Recording of Notes

The investigator begins the process of recording pertinent facts and details of the investigation the moment he arrives at the crime scene. He writes down the identification of the people involved and what he initially observes. He also draws a basic sketch of the crime scene and takes the initial photographs, in order to ensure that a picture of the crime scene is recorded before it can be disturbed. The recording continues for the duration of the crime scene processing.

Searching for Evidence

Each crime scene is different, according to the physical nature of the scene and the crime or offense involved. Consequently, the scene is processed in accordance with prevailing physical properties at the scene and with the need to develop essential evidentiary facts peculiar to the offense. A general survey of the scene is always made, however, noting the location of obvious traces of the action, the probable entry and exit points used by the offender(s), and the size and shape of the area involved.

In rooms, buildings, and small outdoor areas, a systematic clockwise search for evidence is initiated. (A counterclockwise or any other systematic movement may be just as effective in the search. However, in the interest of uniformity, it is recommended that the clockwise movement be used.) The investigator examines each item encountered and the floor, walls, and ceiling to locate anything that may be of evidentiary value. He should give particular attention to fragile evidence that may be destroyed or contaminated if it is not collected when discovered. If any doubt exists as to the value of an item, treat it as evidence until it is proven otherwise.

Ensure that each item or area where latent fingerprints may be present is closely examined and that action is taken to develop the prints.

Carefully protect any impression of evidentiary value in surfaces conducive to making casts or molds. Photograph the impression and then make the cast or mold.

Note stains, spots, and pools of liquid within the scene and treat them as evidence; also note any peculiar odors emanating from the scene. Treat as evidence all other items, such as hairs, fibers, and earth particles, that are foreign to the area in which they are found, as for example scrapings from under victim's fingernails.

Proceed systematically and uninterruptedly to the conclusion of the processing of the scene. The search for evidence is initially completed when, after a thorough examination of the scene, the rough sketch, necessary photographs, and investigative notes have been completed and the investigator has returned to the point from which the search began. Further search may be necessary after the evidence and the statements obtained have been evaluated.

In large outdoor areas, it is advisable to divide the area into strips about 4 feet wide. The investigator searches first the strip on his left as he faces the scene, and then the adjoining strip; this procedure is then repeated until a thorough search has been made of the entire area. Even though an outdoor area considered to be within the scene may be very large and will require considerable time to search, it is imperative that the search be made by investigative personnel. Failure to note and take the proper action on each piece of evidence in this area will be as detrimental to an investigation as it would be in a small room. Several investigators may be utilized to make the search; however, all the recording of locations of items of evidence on the rough sketch should be done by the same investigator, assisted by others as necessary.

It may be advisable to make a search beyond the area considered to be the immediate scene of the incident or crime. For example, evidence may indicate that a weapon or tool used in the crime was discarded or hidden by the offender somewhere within a square-mile area near the scene. Under such circumstances, personnel needed to carry out the search may be secured from another police unit or other available sources. All those participating in the search must be thoroughly briefed on at least the following points:

1. A full description of the item(s) being sought.

2. All information available as to how the item(s) may have been hidden or discarded.

3. The action to be taken when the item is found. The searchers should be emphatically informed that when they discover an item they believe to be the one sought, or one similar to it, they should

immediately notify the investigator in charge of the search, refrain from touching or moving the item, and protect the area until the investigator arrives.

After completing the search of the scene, the investigator should examine the object or person actually attacked by the offender. For example, a ripped safe, a desk drawer that has been pried open, or a showcase from which items have been stolen would be processed after the remainder of the scene has been examined for traces of the offender. In a homicide case, the position of the victim should be outlined with white chalk or any other suitable material before the body is removed from the scene. If the victim has been pronounced dead by a doctor or is obviously dead, it is usually advisable to examine the body, the clothing, and the area under the body after the remainder of the scene has been searched. This practice enables the investigator to evaluate all objects of special interest in the light of all other evidence found at the scene.

Collecting Evidence

Collecting evidence at a scene is usually done after the search has been completed, the rough sketch has been finished, and the photographs have been taken. It may be advisable under certain conditions to collect various fragile items of evidence as they are found. For example, items of evidence that would be destroyed by the elements or become contaminated despite protective measures, and those items that would impede further search, should be collected when they have been located and depicted on the sketch. The essential factor is that evidence be carefully and properly collected.

When collecting evidence, the investigator must handle it as little as possible. Rubber gloves may be used. See the table on p. 6 for recommended methods of handling specific items that may be collected at a scene.

If, during the collection of evidence, the investigator touches a piece of evidence in a manner that leaves his fingerprints on it, he must indicate this fact in his notes and inform the laboratory personnel if they make an examination of the evidence.

It may be necessary to damage, partially destroy, or otherwise decrease the effectiveness of an article in order to collect important evidence—for example, to cut the upholstery on a piece of furniture to obtain an area stained with blood, or to cut out a section of a wall to collect fingerprints that cannot be obtained by other means. Such an action would be based on the merits of the individual case and must have the

Recommended Methods for Handling Specific Items of Evidence

Item	Method
Pistol	Use the fingers on the knurled grips; do not touch smooth metal parts. Use prepared box with a peg for the barrel, or place flat in box for transporting.
Paper, money, documents	Use tweezers. Do not place tweezers over any obvious smudge. Place each item in a clean plastic envelope or bag.
Broken glass	Use the fingers on the edges of larger pieces. Do not touch flat surfaces. Use tweezers on pieces too small for the fingers. Do not grasp over any obvious smudges. Wrap pieces individually in clean tissue and place in a box.
Bottles, jars, drinking glasses	Insert two or more fingers into large-mouth vessels. Place the index fingers on the top and bottom of small-mouth vessels. Do not contaminate or spill any substance in the vessel that may be of evidentiary value.
Bullet	Use fingers or tweezers with taped ends. Avoid damage to rifling marks on the circumference. Place in a pillbox.
Cartridge case	Pick up at the open end with tweezers. Avoid scratching. Place in a pillbox.
Dried stains on a floor	Remove by gouging deeper than the stain with putty knife, wood chisel, or other necessary tool. Place in a pillbox or larger, similar container.
Dried stains on the smooth surface of furniture	Scrape with pocketknife or putty knife, removing as little of the finished surface as possible.

approval of the owner when it has been determined that the action is necessary to the investigation. If a door or window must be removed from a building in order to have it processed at a laboratory or held as evidence, the investigator should ensure that necessary measures are taken to protect the contents of that building.

When collecting evidence at the scene for laboratory analysis, the amounts needed will depend upon the type of evidence and the tests to be conducted. For proper evaluation of stains by laboratory technicians, control samples should be submitted in addition to the collected stains. For example, a stain on soil or porous surfaces is collected by dipping or gouging beneath the stain, and in addition, unstained portions are collected and identified as control samples. The integrity of control samples must be preserved as carefully as that of evidence.

Marking

The investigator places his initials, the date of discovery, and the time on each item of evidence so that it can be identified by him at a later date. These marks should be made as soon after discovery as feasible and in a place least likely to affect the appearance or the monetary or evidentiary value of the item. Evidence that cannot be marked should be placed in a suitable clean container and sealed, and the identifying marks placed on the container. The investigator makes appropriate notes, including a description, in his notebook at the time the evidence is marked.

In instances in which several items with the same appearance are collected, the investigator places an identifying letter on each item and indicates by the letter in his notes and on the sketch where each was found.

Tagging

Physical evidence that the investigator obtains must be tagged prior to its submission to the evidence custodian. This action should take place at the scene of the crime when the evidence is collected, at the place of receipt, or as soon thereafter as possible. The tag serves as an aid in the processing and storage of evidence.

Evaluating

Frequently, the successful conclusion of an investigation depends on an accurate evaluation of the evidence. Each item must be evaluated in relation to all other evidence, individually and collectively. The investigator's evaluation begins with the first information received concerning the incident or crime and continues until the investigation has been satisfactorily concluded or discontinued by proper authority. The evaluation may include a discussion of the evidence with the supervisor, other investigators, laboratory technicians, or other experts in a given field.

Preserving

It is the investigator's responsibility to ensure that every precaution is exercised to preserve physical evidence in the state in which it was received until it is released to the custodian of evidence. Preservation includes security and chain of custody. Usually, a key-type safe is made available to the investigator for temporary storage of evidence during other than normal duty hours.

Releasing Evidence

Evidence, once in the investigator's possession, is released only to the evidence custodian, to another person designated by the investigator's supervisor, or to other competent and legal authority.

Clearing the Scene

Under no circumstances is the scene cleared until all processing has been completed. The clearance should be effected at the earliest practicable time.

SKETCHES AND NOTES

Properly prepared sketches and notes are invaluable to the investigator as a reference when questioning witnesses, suspects, and victims; in preparing a report of investigation; and to refresh his memory when he appears in court. Sketches and notes made during an investigation become the property of the police department and are not retained or used as personal property.

Notes or sketches used to refresh the investigator's memory during a court appearance may be reviewed by the court, so appropriate attention is given to ensure that they are legible and project clear, meaningful facts. Lack of organization in the notes or sketches could unfavorably affect investigative procedures and adversely influence the weight given to the investigator's testimony in a court of law.

Sketches

A sketch graphically portrays the scene of a crime and items within the crime scene that are of interest to the investigation. The sketch, photographs, and investigative notes are mutually complementary, and are all necessary to effectively process the crime scene. The sketch provides the best means of portraying distances between objects at the scene.

To cover items of interest to the investigation, all sketches must, as a minimum, depict:

1. The locations of approaches, such as roadways, paths, entrances, exits, windows, and skylights

2. The size of the area or building

3. The exact locations and relative positions of all pertinent evidence found at the scene

4. The camera location at the time of each photograph of the scene made by the investigator

In depicting the items above, the sketch:

1. Reflects accurate measurements, verified by another person
2. Indicates the compass direction of north
3. Designates the scale (for scale drawing only). If no scale is used, this fact is stated
4. Employs a conventional system of measurement. Paces or steps are not used
5. Contains a legend that explains all symbols or letters used to identify objects on the sketch. Symbols are used where practicable. Lists the report or incident number (if available), offense alleged, name of victim, designation of scene, location of the scene, date and hour the sketch was started, and the name of the person who made the sketch

There are two basic kinds of sketches, the rough sketch and the finished sketch (scale drawing). Both types contain the same general information, but they differ according to the techniques of presenting the information they contain.

Rough Sketch

The rough sketch (Figure 1.1) is drawn at the scene of the crime and is not changed after leaving the scene. It is not usually drawn to scale but depicts accurate distances, dimensions, and proportions through use of one of the methods to be discussed below. The rough sketch is filed with the copy of the report. More than one sketch can be made of a particular scene; it may be necessary, for example, to sketch separately some limited areas within the scene.

Materials needed to prepare a rough sketch are (1) a soft lead pencil (#2), (2) unlined or graph paper, (3) a clipboard, (4) a steel tape (at least 100 feet long), (5) a ruler, and (6) a compass.

Finished Sketch

When possible, the finished sketch (Figure 1.2) is drawn to scale from the information listed on the rough sketch. When drawn to scale, the sketch need not include figures to show distances. If it is not drawn to scale, this fact should be indicated on the sketch and distances should be shown as on a rough sketch. A copy of the finished sketch should be attached to each copy of the report.

10 Crime Scene Processing

The investigator need not prepare the finished sketch, but he must verify its accuracy. It is recommended that the finished sketch be prepared by personnel skilled in such work. The name of the person who prepared it is indicated in the report.

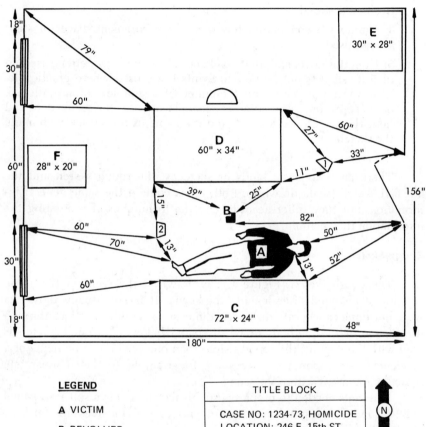

FIGURE 1.1 Rough sketch

Crime Scene Processing 11

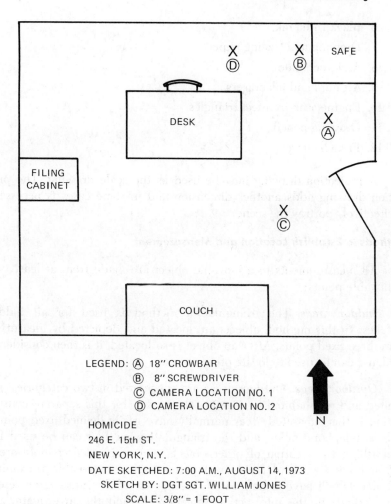

FIGURE 1.2 Finished sketch

If the finished sketch is prepared by personnel of the investigative unit, the following materials are considered necessary:

1. Drawing board or table
2. Draftsman T-square
3. Pencil-and-ink drawing compass
4. Ink-ruling pen
5. Lettering and drawing pens

6. Black India ink
7. Drawing and tracing paper
8. Architect scale
9. Art gum and ink erasers
10. Engineer or architect triangles
11. Drawing pencil
12. French curve

A projection drawing may be used as the scale drawing. The projection drawing adds another dimension and in some cases is necessary to effectively portray the scene.

Methods to Establish Location and Measurement

All measurements to a movable object are made from at least two immovable points:

Indoor areas. The triangulation method is used for all indoor sketches. In this method, objects are located and depicted by measuring from three fixed points. After an object is so located, it is then considered fixed and can be used to locate other objects.

Outdoor areas. Outdoor areas are considered in two categories: inhabited and uninhabited or remote. The reason for this separate consideration is that inhabited areas normally have easily defined fixed points, such as telephone poles, and the triangulation method can be used for establishing the location of other objects. Uninhabited or remote areas may not have easily defined fixed points within reasonable proximity, and objects will have to be located through use of the intersection resection method. In the intersection resection method the investigator requires a compass and a large ratio map of the area. On a large ratio map forms that can be observed by the investigator on the ground can be seen on the map. The map is then matched with the compass to be in relation with the area. The investigator sights or takes a reading on a distant land form. He then gets a second reading on another land form. The two degree readings are drawn on the map. The lines meet at the investigator's position.

Investigative Notes—Basic Principles

Investigative notes are prepared for use in recalling places, events, incidents, or other pertinent facts and are filed with the copy of the report.

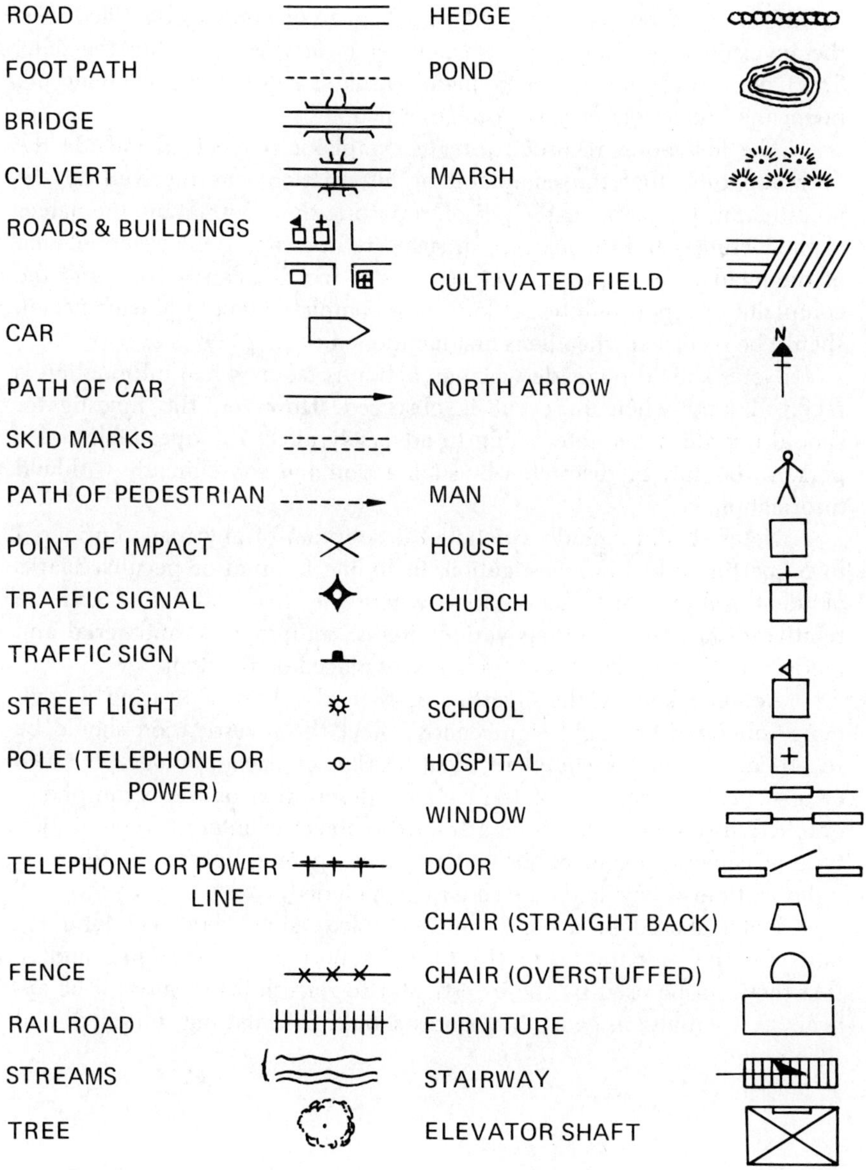

Symbol name		Symbol name	
ROAD	————	HEDGE	⚬⚬⚬⚬⚬⚬⚬
FOOT PATH	‑ ‑ ‑ ‑ ‑ ‑	POND	(pond shape)
BRIDGE	(bridge symbol)		
CULVERT	(culvert symbol)	MARSH	(marsh symbol)
ROADS & BUILDINGS	(buildings symbol)	CULTIVATED FIELD	(field symbol)
CAR	▷		
PATH OF CAR	———▶	NORTH ARROW	↑N
SKID MARKS	= = = = = =		
PATH OF PEDESTRIAN	‑ ‑ ‑ ‑▶	MAN	(stick figure)
POINT OF IMPACT	✕	HOUSE	□
TRAFFIC SIGNAL	◆	CHURCH	(church symbol)
TRAFFIC SIGN	▪		
STREET LIGHT	✸	SCHOOL	(school symbol)
POLE (TELEPHONE OR POWER)	‑o‑	HOSPITAL	⊞
		WINDOW	(window symbol)
TELEPHONE OR POWER LINE	+++	DOOR	(door symbol)
		CHAIR (STRAIGHT BACK)	⌂
FENCE	✕ ✕ ✕	CHAIR (OVERSTUFFED)	⌒
RAILROAD	┼┼┼┼┼┼┼┼	FURNITURE	▭
STREAMS	(stream symbol)	STAIRWAY	(stairway symbol)
TREE	(tree symbol)	ELEVATOR SHAFT	(elevator shaft symbol)

FIGURE 1.3 Drawing and charting symbols for law enforcement officers

Notes should be printed or made in legible handwriting, preferably in ink that does not smudge easily. Each page of notes is identified with the investigator's name, the case number (when known), and the date. Short phrases should be used; single words or shorthand notes may not be meaningful at a later date or to other people.

The first notes recorded after a complaint is received include the date and time that the complaint or information was received by the investigator, the name of the person reporting the information, the names of the victim(s) and the accused or suspected person(s), the exact location of the incident or crime, a brief account of the details received, and the complaint or report number, if known. A complete identity of each person should be recorded when he is first mentioned.

Notes should be made when an action is taken, when information is received, and when an event is observed. However, the investigator should not allow his note taking to adversely affect the questioning of a person who may be distracted by such action and consequently withhold information.

Notes should include a detailed description of any item considered to be pertinent to the investigation, including unusual or peculiar marks of identification; the exact location where the item was found and the relative distances separating various items; trade names, and serial and model numbers; and all identifying marks placed on the item.

Action taken by the investigator that may have a bearing on evidence obtained or might significantly affect the investigation should be recorded, as well as complete identification of each photograph taken of a scene. The position of the camera, description of the scene photographed, distance of the camera from the object of interest, type of film, type of camera, height of the camera, lens opening, length of exposure, lighting, time of day, and date should be included.

Notes should be accurate and complete, since they will form the basis for the preparation of the formal report of investigation, and so that they can be used by the investigator to refresh his memory if he appears as a witness in court. The notes should be lined out, initialed, and then rewritten.

2

Handling and Care of Physical Evidence

INTRODUCTION

This chapter addresses the general precepts of handling and caring for an item of physical evidence after it has been collected, through the time when it is disposed of.

Physical evidence is one of the investigator's most valuable assets in pursuing the investigation to a successful conclusion. It produces leads for him during the conduct of the investigation and aids in establishing the guilt or innocence of an accused person in a court of law.

To achieve the maximum benefit from physical evidence, the investigator must not only be skilled in its collection; he must know how to handle and care for the evidence beyond the time of collection in order to preserve it for the development of leads, for laboratory examination, and/or for presentation in court. Such handling and care involves storing the evidence so as to retain the integrity of the item in its original condition as nearly as possible, maintaining a chain of custody for the item to assure responsibility and to ensure its evidentiary value, its proper transmittal to the laboratory for analysis if necessary, and its disposition when it is no longer of evidentiary value.

DEFINITIONS

Evidence in general is anything that tends to prove or disprove a point under investigation or consideration. *Physical evidence* is that evidence having a physical or material quality—a tangible article, no matter how large or how microscopic.

Physical evidence is divided into two general categories:

1. Movable evidence, which can be picked up at a crime scene or any other location and transported—e.g., tools, weapons, clothing, glass, and documents

2. Fixed or immovable evidence, which cannot be readily removed from a scene because of its size, shape, or makeup—e.g., walls, telephone poles

Fragile evidence is physical evidence that, if special care is not taken to preserve its state, can deteriorate to a point where it is no longer of evidentiary value. It may be difficult to detect; it may be movable or immovable. A footprint in the snow is actually immovable, but a cast of it can be taken and preserved so as to be admissible as evidence. Fingerprints can be "lifted," or removed, whereas body fluids can be preserved in their natural state (or nearly so).

The question frequently arises as to whether an object is or is not evidence. The investigator resolves this question by evaluating the object, the circumstances, and conditions at the scene, supporting his decision with good judgment, common sense, and past experience. If a doubt exists, then the object is secured and processed as evidence. Subsequent evaluation will determine its worth to the investigator and its ultimate disposition.

IDENTIFICATION OF PHYSICAL EVIDENCE

The investigator who first receives, recovers, or discovers physical evidence must be able to identify such evidence positively, at a later date, as being the specific article or item obtained in connection with a specific investigation. Toward this end, the investigator marks and tags the evidence as it is obtained or collected. He should form the habit of marking and tagging evidence promptly.

Marking

Marking is best done by the investigator's inscribing his initials, the date, and the time directly upon individual items of physical evidence. Care must be exercised to place the markings so as not to destroy any latent characteristics on the evidence or to reduce its intrinsic value. When an item cannot be marked without doing so, it is placed in a suitable container, which is then sealed and marked. Evidence such as hair, soil, and fluids cannot be marked, and thus are also placed in a suitable container, sealed, and the container marked.

The use of a Carborundum- or diamond-point pencil is recommended for marking on hard surfaces, and ink on other items. The investigator then records the marking and its location in his notebook.

TABLE 2.1
Methods for Marking Evidence

Item	Method
Pistol, semiautomatic	Use diamond-point or Carborundum-point pencil. Mark on slide, receiver, barrel, and magazine.
Revolver	Use diamond-point or Carborundum-point pencil. Mark on barrel, cylinder, frame.
Bullet	Use diamond-point or Carborundum-point pencil, or hard, sharp-pointed instrument. Mark on base.
Cartridge case, caliber .38 or larger	Use diamond-point or Carborundum-point pencil. Mark just inside the open end.
Cartridge case, smaller than caliber .38	Place in a container and mark container.
Knife	Use diamond-point or Carborundum-point pencil. Mark on the blade as near as possible to the handle.
Liquids	Place in clean glass container (plastic, if the liquid might freeze and break the container) and seal to prevent contamination or leakage. Mark container with diamond-point or Carborundum-point pencil, or attach a label and write the necessary data in ink.
Hairs, fibers, dried blood, and powders	Place in clean pillbox and seal to prevent contamination. Mark container with ink.
Casts of impressions in soil, snow, or other surfaces	Use stick, pencil, or similar marking instrument. Mark on upper surface before cast has hardened.
Handkerchief, towel, flag, or similar item	Use ink. Mark near the edge in an area where there appear to be no deposits of value as evidence.
Coat, dress, or similar item of wearing apparel	Use ink. Mark inside, on a double thickness to lessen the possibility of staining the outer surface.
Glass, other than small fragments, and similar items	Use diamond-point or Carborundum-point pencil, a piece of adhesive tape (appropriately marked), or a grease pencil; mark in area where there appear to be no deposits of value as evidence. Or place in container and mark container with ink.

Tagging

Tagging further serves to help the investigator identify evidence, and facilitates its processing and handling by the evidence custodian and the laboratory technician. The property tag, containing pertinent data about the evidence, is attached to the article or container. It is recommended that the property tag (also known as the "evidence tag") be completed in ink.

CHAIN OF CUSTODY

If it is not properly maintained, an item of evidence may be inadmissible in court. The chain of custody, which ensures continuous accountability, is defined as being made up of all those who have had custody of the evidence since its acquisition by a police agency. It begins when the item is collected, and is maintained until it is disposed of.

Each person in the chain of custody is responsible for the care, safekeeping, and preservation of an item of evidence while it is under his control. Because of the sensitive nature of evidence, an evidence custodian assumes responsibility for the item when it is not in use by the investigating officer or other competent authority involved in the investigation, such as a trial counsel.

Evidence Custodian

Criteria for Selection

The person appointed as evidence custodian should be, if possible, always available to the other investigators within the unit, to receive and release evidence and attend to other administrative matters as required. Ideally, the appointee is assigned to administrative or operations duties, but if operational requirements dictate that a fully committed investigator be appointed as evidence custodian, consideration should be given to appropriately reducing his case load. If necessary, the ranking investigator can act as evidence custodian, but this is undesirable because he should be in a position to supervise the custodian and double-check his procedures.

Alternate Evidence Custodian

The criteria for appointment and the duties of the alternate custodian are the same as those for the custodian. The alternate custodian assists the custodian and is available when the custodian is not; thus the custodian and alternate should not be assigned as an investigative team. The alternate evidence custodian does not make final disposition on any item of evidence.

Records and Forms

Police Receipt for Property

A form is used as the official record of receipt, of chain of custody, and of final disposition of items of physical evidence acquired by criminal investigators.

Evidence Receipt. When this form is used as an evidence receipt, four copies should be prepared. The original and first carbon are presented to the evidence custodian, the second carbon is given to the person from whom the property was received, and the third carbon is placed in the case file.

The receipt contains the following information:

1. Complaint or case number: the police report number or the department (for example, "Homicide") case number, whichever is appropriate.

2. Department designation: the department or unit conducting the investigation.

3. Location: address of the department or unit conducting the investigation.

4. Name of person from whom property is obtained: a block containing the full name and Social Security number. If the property is not received from an individual, or taken from his presence or from a crime scene, this section is marked NA. The investigator should carefully note the circumstances in his investigative notebook.

5. Address: the location where the person who released the property resides and may be reached. If the property is not obtained from a person, or in his presence, this block is also marked NA.

6. Location of property: the exact location of the evidence at the time of acquisition. If the property was recovered from a person, the location should include exactly where on the person the property was found, such as "upper left coat pocket."

7. Purpose for which obtained: normally, "for evaluation as evidence in a criminal investigation."

8. Item number. List each item of evidence in numerical sequence, using arabic numbers 1, 2, 3, etc.

9. Quantity. Use arabic numerals to indicate the quantity of each specific item of evidence acquired.

10. Description of articles. The description of each item of evidence is detailed, accurate, and based upon what is actually observed about the object at the time it is acquired. The item's physical characteristics and condition, especially if it is valuable, are described. The value of articles is *never* estimated or listed, nor the type of metal or stone in jewelry or similar items, beyond a statement of their color, size, and configuration; e.g., an article appearing to be gold is listed as "gold colored metal." Serial numbers are listed whenever possible. The markings used by the investigator to identify the evidence at a later date are also recorded. The words "LAST ITEM," are placed in capital letters after the last item listed. These words are centered on the page and solid lines are drawn in opposite directions from the words to the margin on each side of the form.

Chain of Custody. The item numbers, date of transaction, persons relinquishing and receiving the items, and the purpose for the custody change are filled in appropriately. The first entry under "relinquished by" is the person from whom the property was taken into custody. Should the person refuse or be unable to sign, the following statement is included in the "Description of articles" section: "John Smith refused (was unable) to sign." A witness is then requested to sign under this statement. The first entry under "received by" should be the investigator or other person receiving the property. The purpose for the change of custody should be described briefly, e.g., acquired for evaluation as evidence, delivered to evidence custodian, or transferred to an investigator. This procedure is followed each time there is a custody change. See Figure 2.1 for an example of the chain of custody.

Evidence Voucher. When the original form is presented to the evidence custodian it becomes a voucher and is given a voucher number. Evidence vouchers should be numbered consecutively for each year in the margin at the bottom right corner of the form. The location of the evidence accounted for by the voucher should be penciled in at the bottom left margin of the form; this is erased and changed whenever the location of the evidence changes. For example, this might read, "items one and two in safe, three and four in evidence bin number 6." The chain of custody section on the evidence voucher should be completed whenever any part of the evidence leaves the evidence room or is returned or whenever a new evidence custodian assumes control. The original evi-

ITEM NR	DATE	CHAIN OF CUSTODY		PURPOSE OF CHANGE OF CUSTODY
		RELINQUISHED BY	RECEIVED BY	
2 3	10 Jun 7X	TYPED NAME, GRADE AND BRANCH JERRY T. KILO, Crim Inv SIGNATURE *Jerry T. Kilo*	TYPED NAME, GRADE AND BRANCH Registered Mail # SIGNATURE 2468	Transmitted to Crime Lab for examination
2 3	12 Jun 7X	TYPED NAME, GRADE AND BRANCH Registered Mail # SIGNATURE 2468	TYPED NAME, GRADE AND BRANCH HARRY P. SMOOTH, Lab Tech SIGNATURE *Harry P. Smooth*	Received for chemical analysis
2 3		TYPED NAME, GRADE AND BRANCH HARRY SMOOTH, Lab Tech SIGNATURE *Harry P. Smooth*	TYPED NAME, GRADE AND BRANCH Registered Mail # SIGNATURE 3577	Returned to evidence custodian
2 3		TYPED NAME, GRADE AND BRANCH Registered Mail # SIGNATURE 3577	TYPED NAME, GRADE AND BRANCH JERRY T. KILO, Crim Inv SIGNATURE *Jerry T. Kilo*	Returned from Crime lab

FIGURE 2.1 Evidence subvoucher showing chain of custody

dence voucher does not leave the evidence room except for submission in court, in which case a duplicate copy is maintained in the voucher file to indicate the disposition of the original.

Evidence subvoucher. When only some of the items listed on a voucher are removed from the evidence room, an evidence subvoucher must be prepared. It is made out exactly like the original voucher except that only those items being released are included in the description of evidence.

Evidence Voucher File. This file contains original evidence vouchers and those subvouchers that designate the location of an original voucher. It also contains subvouchers that have been used for interim release of evidence. These are filed with the original voucher to which they pertain.

Evidence Ledger. The evidence ledger provides a second method of accountability for evidence, the voucher file being the first. Entries are made in ink and are separated by a line drawn across both pages of the ledger. The descriptions of the articles of evidence do not need to be as lengthy as on the evidence voucher.

Evidence Depositories

The sensitive nature of evidence requires that it be secured at all times. Normally, evidence is stored in a room designated for that purpose, unless circumstances dictate that a temporary depository be used. Some important general requirements are enumerated below.

Temporary Depositories. A safe or filing cabinet suitable for storing material may be used for retaining small items of evidence temporarily. Evidence may also be stored temporarily in a building or enclosure not meeting the normal standards of an evidence room, if appropriate security measures are taken.

Evidence Room. The basic requirements for an evidence room are as follows: The room must lend itself to meet the physical requirements set forth by the local police department standards (and often, state requirements). The design of the room should allow construction of bins, shelves, and cabinets that can be secured with lock and key for additional security. The overall capacity of the room must be adequate to accommodate the normal amount of evidence handled by the department involved.

Internal fixtures necessary to the evidence room include a refrigerator for perishable or unstable items; shelves or bins for orderly arrangement of appropriate items of evidence; a desk or table for the evidence custodian and a file cabinet for evidence files, the size dependent on the amount of evidence handled; and safes or filing cabinets, not smaller than a field safe, one to be used for items of evidentiary value, the other for marihuana, narcotics, and dangerous drugs.

Combination and Key Control

Combinations and keys to both temporary depositories and evidence rooms are retained only by the custodian and his alternate. Combinations are changed upon change of either the primary or the alternate custodian. Each time combinations are changed, they should be recorded, placed in a sealed envelope, and maintained in the department safe. For each lock, two keys should be maintained, one by the evidence custodian. Duplicate keys should be placed in separate sealed envelopes and kept in the department safe, under the direct control of the chief of police or his designated representative.

Inventories and Inspections

All inventories conducted are made a matter of written record. Through such inventories, errors in procedure or loss may be discovered before they become too grave. The only effective way to ensure against errors and losses, however, is the effective execution of duties by the evidence custodian. His actions directly affect the admissibility of evidence in court.

Inventories

Monthly. The evidence custodian should inventory the evidence depository every 30 days. He verifies the evidence in the depository against evidence vouchers and the evidence ledger, ensuring that all postings are current as of the previous inventory.

Quarterly. A disinterested officer should inventory the evidence depository once each calendar quarter, using the same procedures mentioned under the monthly inventory.

Upon Change of Custodian. A joint inventory should be conducted by the incoming and outgoing custodians when there is a change of custodians. When the incoming custodian is satisfied that all is in order, he signs the "Received" block on all current evidence vouchers, thereby assuming responsibility.

Inspections

The commanding officer, or the officer exercising supervision over the evidence custodian, should inspect at least every 30 days to ensure that the evidence depository meets specified standards. Informal inspections should also be made from time to time by the commander, to maintain the proper degree of supervision on a continuing basis.

DISPOSITION OF EVIDENCE

When an article is no longer of evidentiary value, it is disposed of in accordance with departmental procedures. The evidence custodian is responsible for constantly monitoring evidence on hand to ensure that proper and timely disposition is made. To determine when an item of evidence should be disposed of, the evidence custodian consults with the investigator who originally procured it, and any other investigator who has an official interest, to make sure the item is no longer needed as evidence.

An item no longer of evidentiary value is returned to the owner if ownership is known. Where there is no conclusive proof or judicial determination of ownership, the item is returned to the person from whom it was taken. If judicial proceedings concerning ownership are pending, the property is held by the evidence custodian until the findings of such proceedings are known. An item for which an owner cannot be located is disposed of as lost, found, or abandoned property.

Evidence released to trial counsel temporarily for presentation in court will be returned to the evidence custodian for safekeeping until final disposition can be made. The custodian does not dispose of such evidence until all court appeals have been completed, or until directed by the appropriate departmental authority.

A large item or a smaller serial-numbered item of evidence that is necessary to the operation of a unit or the daily life of the owner, and the value of which as evidence cannot be influenced by repair or alteration, as with bodywork on a car fender, may be released to the appropriate unit or owner after it has been properly recorded as evidence. The evidence custodian should not, however, hesitate to hold any item of monetary value whose evidentiary value could be destroyed if released, including those items mentioned above.

Contraband is released to the appropriate federal agency, except in the case of marihuana, narcotics, and dangerous drugs. Weapons that are in violation of federal law are released to the local agent of the Alcohol, Tobacco and Firearms Bureau of the Treasury Department. Counterfeit currency and counterfeiting equipment are released to the Secret Service Bureau of the Treasury Department. Narcotics, marihuana, and dangerous drugs, when no longer needed for legal proceedings, are destroyed by the custodian in the presence of witnesses or, if requested, released to the Drug Enforcement Administration.

A statement of destruction should be prepared to record the destruction of any evidence. The original is filed with the appropriate evidence voucher.

PACKAGING AND TRANSMITTING EVIDENCE

Packing and Wrapping

Evidence should be packed and wrapped in a manner that minimizes friction and prevents it from shifting, breaking, leaking, or contacting other evidence.

Items such as glass, glass fragments, evidence in glass containers, impressions, casts, ammunition, bullets, and cartridge cases, which are particularly susceptible to breaking, marring, or other destructive changes, should be packed in cotton or soft paper.

When evidence is to be examined for fingerprints, each item should be packed in a manner that prevents damage to the prints. This is done by fastening the object in a container so that it will not shift and so that no other object will come in contact with the area suspected of containing fingerprints.

Liquid evidence, with the exception of explosives, oils, and gasoline, should be packed in all-glass, sterile bottles or other containers and sealed with wax or other suitable materials.

In general, small, solid items, such as bullets, fibers, hairs, paint scrapings, powder and powder patterns, and threads, should be placed upon a piece of plain paper, and the paper folded around it, packed in separate pill or powder boxes, paper containers, or druggist folds, and the packages sealed with adhesive tape, wax, or other suitable material.

Documents, exemplars, standards, string, twine, and rope should be placed in an inner cellophane envelope and an outer manila envelope. Cellophane is not suitable for packing any item that will rust or corrode.

Packages containing acids, ammunition, alkalis, gasoline, glass fragments, guns, liquids, matches, medicines, chemicals, drugs, paints, or other items that require careful or selective handling while in transit should be labeled "Corrosive," "Explosives," "Firearms," "Fragile," "Gasoline," "Keep away from fire," or "Keep cool," as appropriate. Materials bearing traces of accelerants like those recovered in arson should be sealed in either a metal or a glass container, such as a mason jar. They should not be sent in plastic bags, because accelerants will leak through plastic.

The completed packing and wrapping will generally require a combination of the methods indicated above. The exact procedure to be used will depend on the item to be submitted, its quantity, condition, and size, and the method of transmittal.

Transmitting Evidence to the Laboratory

Methods of Transmittal

Four methods by which evidence may be transmitted to a crime laboratory are air express, courier, railroad express, and registered mail. The selection of the method of transmittal will be influenced by the type of evidence to be transmitted and the urgency with which the results of a laboratory examination are needed. The ideal means is for the investigator requesting the examination to hand-carry the evidence to the laboratory.

Chemicals, gases, unexploded bombs, detonators, fuses, blasting caps, and other explosive or inflammable materials will *not* be transmitted through the mail. The transmittal of these items of evidence must conform to the provisions of Interstate Commerce Commission regulations, and appropriate state and municipal ordinances. *Before such items are forwarded, the laboratory must be notified that the shipment is planned and should acknowledge receipt of notification.* Include in the notification the details of how the materials are packed, in order to minimize the danger of unpacking at the laboratory.

A package wrapped for shipment to the laboratory consists of evidence from only one investigation. Violation of this procedure could result in cross-contamination of certain types of evidence and possible problems in chain of custody.

Laboratory Examination Request

All evidence is accompanied to the laboratory by a laboratory examination request. The request (original copy) and the evidence voucher are placed in an envelope. In the case of transmittal by registered mail, the envelope is taped to the outside of the box. Evidence should not be sent in envelopes, because this increases the possibility of damage or contamination. Evidence sent by air express or railroad express should have the envelope inside the box. The address should contain an attention line to a specific laboratory section, if possible, such as "Firearms Section."

PACKING AND TRANSMITTAL CHART

Table 2.2, which follows, is not all-inclusive. If the evidence to be submitted is not listed, consult the specimen list for an item similar in nature and submit accordingly, or contact the laboratory for instructions.

Federal laws prohibit transmitting certain types of merchandise through postal channels. If there is any question of mailing, the nearest postmaster should be consulted. The nature of the evidence will govern the use of warning notices to be affixed to the outside wrapper or box, such as "fragile," "expedite," "corrosive," and so on.

If feasible, the package should be mailed or expressed to a specific section of the laboratory. Whenever possible, evidence should be hand-carried to the laboratory, preferably by the investigator concerned with the investigation. If it cannot be hand-carried, it is forwarded by the means shown in the "transmittal" column.

Items containing stains, such as clothing with stains of blood or other body fluids, should *not* be placed in airtight plastic containers. "Sweating" and moisture accumulation may result within such containers, contaminating the evidence. Such items must be dry before packaging.

Packaging materials such as ice-cream boxes, pillboxes, and plastic bags should be free from all contamination and preferably new and unused.

TABLE 2.2
Packing and Transmitting Physical Evidence to the Criminal Investigation Laboratory

Specimen	Identification	Amount Desired		Preservation	Wrapping and Packing	Transmittal	Miscellaneous
		Standard	Evidence				
Abrasives, including Carborundum, emery, sand, metal filings, etc.	Label and tag on outside of container. Show type of material, date obtained, investigator's name or initials, case, and number.	Not less than 1 ounce and up to 1 pound	All	None	Use containers such as ice-cream box, metal pillbox or powder box. Seal to prevent any loss.	Registered airmail, United Parcel Service (UPS) or air express	
Acids	"	One pint	All to 1 quart, but not less than 15 cc. (½ oz) if available	"	All-glass bottle. Tape in stopper and seal with paraffin wax or plaster of paris. Pack in sawdust, glass wool, or rock wool. *Use bakelite- or paraffin-lined bottle for hydrofluoric acid.*	UPS only	Label ACIDS, GLASS, CORROSIVE

TABLE 2.2 continued

Specimen	Identification	Amount Desired		Preservation	Wrapping and Packing	Transmittal	Miscellaneous
		Standard	Evidence				
Adhesive tape	Label and tag on outside of container. Show type of material, date obtained, investigator's name or initials, case, and number.	One foot	All	None	Place on waxed paper or cellophane. Pack in pill or powder box, paper container, or druggist's fold. Seal edges.	Registered airmail	
Alkalis (Caustic soda, potash, ammonia, etc.)	"	One quart liquid, 1 pound solid	All to 1 qt. All to 1 pt. Not less than 15cc. or ½ oz. if available.	"	Glass bottle with rubber stopper held in with adhesive tape. Seal with paraffin wax.	UPS only	Label ALKALI, GLASS, CORROSIVE
Ammunition	"	Two	Up to five rds. plus specifications and lot number if available	"	Pack in cotton, soft paper, or cloth in small container. Prevent friction, shifting, and contact while in transit. Place in wooden box.	"	If standard make, usually not necessary to send. Label EXPLOSIVE.

Specimen	Identification	Amount Desired	Standard of Comparison	Precautions	Wrapping and Packing	Method of Transmittal	Remarks
Anonymous letter, extortion letters	Place in cellophane envelope, seal with sticker, place date and investigator's initials on sticker. (Make detailed notes describing letter in notebook.)	All	Submit as much as possible of suspect's handwriting, including words in questioned document if possible. Submit fingerprint cards for all persons known to have handled letter.	Do not handle with bare hands.	Cellophane envelope placed in manila envelope; insert stiff backing to prevent bending or folding, seal and mark as for identification. Wrap securely. If burned and/or brittle, obtain instructions from laboratory.	Registered airmail	Include original envelope. Advise if letter should be treated for latent fingerprints.
Blasting caps	Label and tag on outside of container. Show type of material, date obtained, and investigator's name or initials.	All			Should not be forwarded until advised to do so by the laboratory. Packing instructions will be given at that time.		
Blood: *a.* Liquid	Use adhesive tape on outside of test tube. Name of victim or subject, date taken, doctor's name, investigator's name, case, and number.	⅙ ounce (5 cc.) collected in a sterile test tube or Sheppard (Vac) tube		Sterile tube only. NO PRESERVATIVE. NO REFRIGERANT.	Wrap in cotton, soft paper. Place in mailing tube or suitably strong mailing carton to prevent breakage and spillage.	Airmail special delivery, registered	

TABLE 2.2 continued

Specimen	Identification	Amount Desired		Preservation	Wrapping and Packing	Transmittal	Miscellaneous
		Standard	Evidence				
b. Drowning cases	"	Two specimens, one from each side of heart		"	"	"	Consult laboratory if preservative is required.
c. Small quantities:							
(1) Liquid	"	See a above.	All, to ⅙ oz. 5cc.	"	"	"	Collect by using eyedropper or clean spoon, transfer to sterile and chemically clean test tube.
(2) Dry stains	On outside of metal pillbox, powder box, or druggist fold. Type of specimen, date collected.	See a above. Also, control specimen of material (soil, porous matter, etc.) from which stain collected.	As much as possible	Keep dry	Tops, ends, and all folds sealed to prevent leakage	Registered airmail	Dry completely under natural conditions.
d. Stained clothing, fabric, etc.	Using string tag and/or mark directly on clothes. Type of specimens, date secured, investigator's name, case, and number.	See a above.	As found	If wet when found, dry under natural conditions. USE NO EXCESSIVE HEAT TO DRY. No preservative.	Each article wrapped separately and identified on outside of package. Place in strong box, packed to prevent shifting of contents.	Registered airmail, UPS or air express	Submit not less than 1 square inch of stain and 1 square inch of unstained material.

	Identification mark	Standard	Wrapping and packing	Transmittal	Remarks
Bullets (not cartridges)	On base. Investigator's initials and date, or other individual identifying character.	None	Place on cotton or soft paper. Place in pill, powder, or match box. Pack to prevent shifting in transit.	Registered airmail	
Cartridges or rounds (live)	On outside of case near bullet end. Investigator's initial or other individual identifying character.	Two	"	UPS	
Cartridges (empty shells)	Preferably on inside near open end, or outside near open end. Investigator's initial and date, or other individual identifying character.	All	"	"	
Cartridges (empty shells)	Preferably on inside near open end, or outside near open end. Investigator's initial and date, or other individual identifying character.	All found	"	Registered airmail	When fingerprint evidence possible, place in test tube, seal, and label.
Casts (see Impressions)					
Charred or burned paper	On outside of container, indicate type of material, date obtained, investigator's name or initials, case, and number.	All	Pack in rigid container between layers of cotton. If fragile and brittle, consult laboratory for shipping instructions.	Registered airmail	Added moisture, with atomizer or otherwise, not recommended.

TABLE 2.2 continued

Specimen	Identification	Amount Desired		Preservation	Wrapping and Packing	Transmittal	Miscellaneous
		Standard	Evidence				
Checks (fraudulent)	See Anonymous letter.	See Anonymous letter.	"	"	Wrap securely. Also see Anonymous letters.	"	Advise what parts questioned or known.
Check protector, rubber stamp, and date-stamp sets, known standards	Place investigator's name or initials, date, name of make and model, etc., on sample impressions.	Obtain several copies in full word-for-word order of each questioned check-writer impression. If unable to forward rubber stamps, prepare numerous samples with different degrees of pressure.	"	"	Wrap securely, prevent shifting or damage while in transit. For transmitting standards, see Anonymous letter.	"	Do not change the ribbon or alter the inking. Also see Typewriting.
Clothing	Attach property tag and mark directly on clothing the type of evidence, investigator's name, date, case, and number.		"	"	Each article individually wrapped, with identification written on outside of package. Place in strong container.	Registered airmail, or UPS or air express	Leave clothing whole. Do not cut out stains. *If wet, dry before packing*

32

Codes and ciphers	As Anonymous letters	See Anonymous letter.	"	As Anonymous letters	See Anonymous letter.
Documents:					
a. Questioned and secret writing, handwriting and printed specimens, fraudulent checks, anonymous letters, extortion notes, codes, ciphers	As Anonymous letters. Also, obtain statement from suspect acknowledging exemplars as his own and given voluntarily.	"	"	"	See Anonymous letter. (Also see Handwriting footnote.)
b. Check protectors, rubber stamps, datestamp set	See Check protector.	See Check protector.	"		
c. Typewritten material	See typewriting.	See Typewriting.	"		
Drugs:					
a. Liquids	Label or mark bottle in which found with investigator's name, date, case, and number.	All to 1 pt. Not less than 15cc., or ½ oz., if available.	"	If bottle has no stopper, transfer liquid contents to glass stoppered bottle and seal with adhesive tape and wax.	Registered airmail, or UPS or air express.	Mark FRAGILE Determine alleged normal use of drug and, if prescription, check with druggist to determine supposed ingredients.

TABLE 2.2 continued

Specimen	Identification	Amount Desired		Preservation	Wrapping and Packing	Transmittal	Miscellaneous
		Standard	Evidence				
b. Powders, pills, and solids	Label or mark outside of metal pillbox, investigator's name, date, case, and number.		All to ¼ pound	None	Seal with tape to prevent any loss.	Registered airmail, or UPS or air express	Determine alleged use and prescription data as above.
Dynamite and other explosives	Consult the laboratory and follow their telephonic or telegraphic instructions.		All	Should not be forwarded until advised to do so by the laboratory. Packing instructions, if any, will be given at that time.			
Fibers (see Hair and/or Rope)							
Firearms	Attach property tag with pertinent data. Investigator's initials and date on barrel, slide, and frame. Record data in investigator's notebook.			Keep from rusting	Wrap in paper and identify contents of package. Place in cardboard or wooden box. Prevent shifting while in transit. Label FIREARMS.	Under 4 pounds, registered airmail. Over 4 pounds, UPS	Unload all weapons before shipping. Advise if firearms are to be examined for fingerprints.

Fuse, safety	Attach property tag and/or gummed paper label, with investigator's name, date, case, and number.	One foot	None	Place in manila envelope, box, or other suitable container.	Registered airmail, or UPS or air express
Gasoline	Label or mark outside of all-metal container; indicate type of material, investigator's name, date, case, and number.	One quart	Fireproof container	Metal container packed in wooden box.	UPS only	Label GASOLINE.
Glass fragments	Adhesive tape on each piece. Investigator's name, date, case, and number on tape. Separate the questioned and the known.	All	Submit fingerprint cards for all persons known to have handled glass.	Avoid chipping.	Wrap each piece separately in cotton. Pack in strong box to prevent shifting and breakage. Identify contents.	Registered airmail, UPS or air express	Mark FRAGILE.
Guns (see Firearms)					
Gunpowder on cloth	Attach property tag and mark directly on article, type of material, date, investigator's name, case, and number.	All to one gallon	None	Fold fabric flat and then wrap so that no residue is lost through friction. Place clean paper between folds.	Registered airmail	Avoid shaking.

35

TABLE 2.2 continued

Specimen	Identification	Amount Desired — Standard	Amount Desired — Evidence	Preservation	Wrapping and Packing	Transmittal	Miscellaneous
Hair and fibers	Label or mark outside of container, type of material, date, investigator's name, case, and number.	About 20 pulled hairs. Must be representative specimens from different parts of head or body. Keep hairs from various parts separate.	All	"	Druggist fold in metal pillbox. Seal edges and openings with cellophane tape or adhesive tape and also seal container.	"	Envelope not satisfactory.
Handwriting* Handprinting and forgeries, known standards or exemplars (see Documents.)							
Impressions: Plaster casts, tire treads, footprints.	On back, before plaster hardens, place investigator's initials and date.		Up to 2 ft.	"	Wrap each cast in soft paper or cotton, surround with packing material in box to prevent shifting or breakage.	"	Label FRAGILE.

Item						
Matches	Mark or label outside of container, type of material, date, investigator's name, case, and number.	One to two books of paper matches. One full box of wooden matches.	Up to ¼ lb.	Keep away from fire.	Metal container, packed in larger package to prevent shifting. Matches in box or metal container.	UPS or registered airmail	Label KEEP AWAY FROM FIRE.
Medicines (see Drugs)						
Metal	Label or mark outside of container, type of material, date, investigator's name, case, and number.	One pound or 1 foot	All, to 1 pound or 1 foot	Keep from rusting.	Wrap in paper if solid. Use paper boxes or containers if filings. Seal and use strong paper or wooden box.	Registered airmail, UPS or air express	Melt number, heat treatment, and other specifications of foundry, if available
Oil	"	One quart, together with specifications	All to 1 quart. Not less than 1 pint, if available.	Keep away from fire.	Metal container with tight screw top. Pack in strong box using excelsior or similar material.	UPS only	DO NOT USE DIRT FOR PACKING MATERIAL.

*Reproduce the original writing conditions as to speed, slant, size of paper, size of writing instruments, etc. Do not allow suspect to see questioned writing. Give no instructions as to spelling, punctuation, etc. Remove each sample from sight as soon as completed. Suspect should fill out blank check forms in check cases. In handprinting cases, submit 20 to 50 specimens; both upper- and lowercase samples should be obtained. In forgery cases, sample signatures of the person whose name is forged should be forwarded. Submit 15 to 20 specimens if the writing is merely a signature. If writing contains 20 or more words, submit five or more specimens.

TABLE 2.2 continued

Specimen	Identification	Amount Desired		Preservation	Wrapping and Packing	Transmittal	Miscellaneous
		Standard	Evidence				
Organs of body	Label or mark outside of container, victim's name, date of death, date of autopsy, name of doctor, investigator's name, case, and number.		All, to 1 pound. Consult with laboratory.	Consult laboratory. Dry ice in package not touching glass jars.	All-glass containers (glass jar with glass top). Seal with wax. Pack in strong box using excelsior or similar material.	UPS or air express	Label FRAGILE. Keep cool. Metal-top containers must not be used. Send autopsy report.
Paint:							
a. Liquid	Label or mark outside of container, type of material, origin if known, date, investigator's name, and number.	One-fourth pt.	All, to ¼ pt. Not less than 30 ccs., or 1 oz., if available.	None	Friction-top paint can or large-mouth screw-top jars. If glass, pack to prevent breakage. Use heavy corrugated paper or wooden box.	Registered airmail, UPS or air express	
b. Solid	Label or mark outside of container, type of material, date, investigator's name, case, and number.	At least ½ sq. inch of solid	All. If on small object, send object.	Wrap so as to protect smear.	If small amount, pillbox or small glass vial with screw top. Seal to prevent leakage. Envelopes not satisfactory.	Registered airmail or UPS or air express.	Do not pack paint chips in cotton or secure with cellophane or adhesive tape.

Plaster casts (see Impressions).			
Poison (see Drugs)			
Powder patterns (see Gunpowder)			
Powders (see Drugs)			
Rope, twine, and cordage	Tag and/or mark container, type of material, date, investigator's name, case, and number.	Two feet or 1 pound	All up to 2 feet	Wrap securely in clean paper. If strands or fibers, use druggist fold in pillbox. Seal edges and openings with cellophane or adhesive tape.	Registered airmail
Safe insulation or soil	Label or mark outside of container, type of material, date, investigator's name, case, and number.	½ pound	All, to 1 pound. Not less than ¼ pound, if available.	Use containers such as icecream box, pillbox, or powder box. Seal edges and corners to prevent any loss.	Registered airmail, or UPS or air express	Avoid use of glass containers.

TABLE 2.2 continued

Specimen	Identification	Amount Desired		Preservation	Wrapping and Packing	Transmittal	Miscellaneous
		Standard	Evidence				
Semen stains	Tag and/or mark article, type of material, date, investigator's name, case, and number.		Entire article	Avoid friction with stained area.	Fold carefully, protect area with clean paper. Pack to prevent shifting in transit.	Registered airmail	Do not roll item. Do not fold or crease stained area.
Tools	Label on tools and/or property tag, type of tool, identifying number, investigator's name, case, and number.		All		Wrap each tool in paper. Use strong cardboard or wooden box with tools packed to prevent shifting.	Registered airmail, or UPS or air express	
Toolmarks	On object or on tag attached to it, or on opposite end from where toolmarks appear. Investigator's name and date.	Send in the tool.	"	Cover ends bearing toolmarks with soft paper and wrap with strong paper to protect ends.	After marks have been protected, wrap in strong wrapping paper, place in strong box and pack to prevent shifting.	"	Keep questioned specimens separate from known standards.

Typewriting	Place investigator's name or initials, date, serial number, name of make and model, etc., on same side as samples of typewriting.	Obtain at least one copy in full of word-for-word order of questioned typewriting. Also include partial copies in light, medium, and heavy degrees of touch. Also standard and carbon-paper samples of every upper- and lowercase character on the keyboard.	All	None	Wrap securely. Also see Anonymous letter.	Registered airmail	Examine ribbon for evidence of questioned message thereon. For carbon-paper samples, either remove ribbon or place in stencil position.
Urine or water	Label or mark outside of container, type of material, name of subject, date taken, investigator's name, case, and number.	Preferably all urine voided over a period of 24 hours, preferably not less than 30cc., or 1 oz., per specimen.		None. Use any clean bottle with leakproof stopper.	Bottle sealed and packed, surrounded with absorbent material to prevent breakage. Strong cardboard or wooden box.	"	Keep samples separate.
Wire (Also see Toolmarks)	On label or tag. Type of material, date, investigator's name, case, and number.	One foot	All	"	Wrap securely, pack to prevent friction, shifting, breakage, or contact while in transit.	"	Keep questioned specimens separate from known standards.
Wood	"	"	"	"	"	"	"

TABLE 2.3
Standard Military and Commercial Explosives

Type	Characteristics	Trade names—common uses	Results of detonation
LOW-ORDER EXPLOSIVES	SLOW RATE OF CHANGE TO GASEOUS STATE. PUSHING RATHER THAN SHATTERING EXPLOSIVES EFFECT. TWISTING AND TEARING TYPE OF DEFORMATION. COMMERCIALLY USED AS PROPELLING CHARGE FOR MOST TYPES OF SMALL-CALIBER AMMUNITION		
1. Black powder	Low speed of burning. Detonated by safety fuse, flame, or shock.	When mixed with potassium nitrate, is called "US Army Black Powder."	Standard gunpowder prior to introduction of smokeless powder. Now used as expelling charge for shrapnel shells, primers, safety fuses. Leaves bluish-grey residue, greyish smoke.
2. Smokeless powder	Grained into solid cylindrical or cylindrical perforated particles. Coated with graphite.		Universal propellant for standard service ammunition. Detonated by shock, explosion is relatively slow, giving strong propelling force, but little danger of bursting gun barrel.
3. Gas explosions	Usually accidental, but may be result of criminal or subversive acts.	Commercial and industrial gases. Illuminating gas, propane, butane, carbon monoxide, natural gas.	Strong odor of gas, low boom sound. No definite, localized point of origin of force. Velocity of shock wave is low. Injuries mainly from falling debris.
4. Dust or grain explosions	Usually accidental, as result of spontaneous combustion ignited by sparks or friction.		
Volatile Liquids and Incendiaries	Chemical agents that cause ignition of combustible substances by generation of heat or flame		

5. Magnesium | Silver-white metal, malleable into powdered or flake form. | Used as filler or as casing in incendiary shells and bombs. Used for signaling and illuminating flares. | Burns readily with brilliant, bluish-white light and evolution of great heat of combustion.

6. Phosphorous White (chemical warfare symbol WP) | Pale-yellowish, translucent solid. Spontaneously inflammable at normal temperatures. | Powerful incendiary for use in bombs, shells. Screening smoke in hand grenades and mortars. | Gives off dense white smoke. Burning pieces adhere to skin and clothing.

7. Thermit (chemical warfare symbol TH). | Composed of aluminum powder and iron oxide. Reaction initiated by strong heat proceeds throughout mixture. | Used alone as incendiary agent or as primary igniting incendiary material, or for igniting magnesium-cased incendiary bombs. | Evolution of great heat and formation of molten iron and slag.

HIGH-ORDER EXPLOSIVES

EXPLOSION TRANSMITTED INSTANTANEOUSLY THROUGHOUT THE MASS TO PROVIDE DETONATION EFFECT. RAPID CHANGE TO GASEOUS STATE. SHATTERING AND FRAGMENTATION NEAR EXPLOSION CENTER. DEBRIS AND HIGH-VELOCITY FRAGMENTS IN RADIATING PATTERN FROM EXPLOSION CENTER.

Primary high explosives | | | Extremely sensitive to heat, shock, friction. Will not ordinarily burn. Usually detonated when ignited by fuses, heated wires, firing pins. Used only in primer and detonator mixtures because of inferior explosive effect. Used to initiate explosion of blasting, propellant, and bursting charges at desired time. Strength of explosive effect is inferior, but sufficient generation of heat and shock to detonate secondary high explosives.

1. Mercury fulminate | White or grey crystals. Sensitive to flame, impact. Must be kept moist until used. | Initiator or primer for bringing about detonation of high explosives or ignition of powder.

2. Lead azide | White crystalline powder. Very stable initiation, insensitive to fuse or flame. | Used in loading fuse detonators and priming compositions. Usually used only in aluminum detonators.

Secondary High Explosives | | | Insensitive to shock, friction, and heat. Massive explosive effect in relation to size of explosive charge. Used for boosters and as bursting charges. Standard filler for bombs, shells, torpedoes, etc.

3. Tetryl | Yellow crystalline solid. Sensitive to friction, shock, and sparks when confined. Used in form of pressed pellets. | Standard booster explosive. Violence of detonation insures complete, high-order detonation of bursting charge. | Produces great fragmentation when used in heavy artillery shells.

TABLE 2.3 continued

Type	Characteristics	Trade names—common uses	Results of detonation
4. TNT (trinitrotoluene)	Melted for pouring into bombs or shells. Safety of storage, manufacture, transportation. Insensitive to blows or friction when not confined. Requires booster charge when in cast form.	Triton, Trotyl, Trilite, Trinol, Tritilo. Bursting charge for high-explosive shells, bombs, mines, torpedoes. All types of blasting; demolition of bridges, railroads, fortifications. Usually used in compressed block form for blasting purposes.	Powerful explosive effect. Black smoke, slight suffocating odor.
5. RDX explosives	Synthetically manufactured. Combined with Vaseline or mineral oils to make "plastic explosives." Ease of molding, high-speed detonation. Insensitive to cold, moisture, shock.	Cyclonite, C4, Onit. Used in demolition work.	Heavy black smoke. Suffocating odor occasionally present.
6. PETN	Usually in pressed form. Extremely sensitive, will detonate from impact of rifle bullet. White in pure state. Pressed form is rose color, can be broken, cut, or bored.	Use is restricted because of its sensitivity. Used for boosters and bursting charges in small-caliber ammunition. Used as base charge in bombs, sea mines, torpedoes.	Greyish-black smoke. Slight "gunpowder" odor.
7. Blasting gelatine	Most powerful explosive used industrially. Amber yellow color, rubbery consistency.	Torpedo Explosive No. 1, Sprenggelatine, Nitroglycerin Gelatine, Gummidynamite. Strongest, quickest water-proof explosive. Used in special cases of tunnel-driving, shaft sinking, deep-well shooting, submarine work.	Slight aromatic odor. Dust, sand, powdered stone results from detonation on land.

8. Nitroglycerin	Thick, oily, greenish-brown liquid. Insoluble; cannot be destroyed by pouring in water, down drain, or on ground. Detonated by shock or contact with metal. Destroyed only by burning or detonation.	Too sensitive to be used alone. Made safe for commercial use by absorbing it into infusorial earth to make dynamites and blasting gelatines.	Violent but localized reaction. Difficult to control extent of explosion when detonated in liquid pure form.
9. Powder explosives (non-gelatinous)	High detonation speeds whether confined or not. Sensitive to water and dampness. Require detonators.	Trojan Explosive, Trojan Coal Powder, Big Red, Apcol, Black Diamond, Carlsonit, Antracite, Austin Red Diamond, Apache. Coal powder. Limited commercial and military use.	Bursting explosive effect.
10. Straight dynamite (non-gelatinous).	Contains 15 percent to 60 percent nitroglycerin in absorbent materials. Made into brownish powder or puttylike material and cast into familiar stick form. Sensitive to shock, friction, heat. Fast, shattering, but localized explosive effect. Most water-resistant form of dynamite.	Ajax Powder, Polar Ajax, Polar Samsonite, Low-Freezing Dynamite, Forcites. Chief commercial blasting explosive. Main military use is for demolition and scrapping work, submarine blasting. Used in ditch blasting.	Comparable in strength to TNT, but has less shattering effect. Obnoxious fumes. Confined explosive effect allows use for blasting in coal mines having inflammable atmosphere of gas–air or coal-dust–air mixtures.
11. Gelatine dynamite	Composed of gelatinized nitroglycerin absorbed in vegetable meal. Puttylike material is cast into stick form or left in plastic consistency. Waterproof. Plastic form is most familiar commercial type.	Gelignite, Grisoutine, Grisoutite. Used in hard, tight work where maximum shattering effect is desired. Can be loaded solidly in bore holes.	Greyish-yellow smoke. Strong bitter-almond odor. More shattering effect than straight dynamite.

3

Trace Evidence

INTRODUCTION

This chapter acquaints the criminal investigator with various techniques for finding and handling trace evidence during a criminal investigation, and with the services provided by the police crime laboratory in examining, comparing, and evaluating such evidence. It does not cover all material items that could be classified as trace evidence, nor does it cover all those investigative techniques that may be used for finding and handling such evidence. It is intended that the points mentioned will alert the investigator to other possibilities, and by association, suggest the further development and perfection of investigative skills.

Trace evidence at a crime scene can include obvious items, such as bloodstains; or inconspicuous items, such as dust particles, that are easily overlooked, frequently mishandled, and all too often discarded as outright useless. Police files contain numerous case histories emphasizing the value of properly handling trace evidence in proving a case against a suspect or clearing an innocent person.

The investigator must be alert to the consequences of improper handling of trace evidence, particularly since it may completely negate the value of otherwise admissible court evidence. For example, if a suspect is returned to the scene of a crime before the scene is completely processed, he could claim that the hairs found there were left during the return visit. The same assertion may diminish or negate the value of other trace materials found at the scene, such as paint chips, fibers, dirt, and so on. The thoughtless intermingling of trace evidence found at different parts of the crime scene may also render valuable evidence worthless. For these reasons, the investigator must always observe the cardinal rule for handling physical evidence, particularly trace evidence: *Avoid contamination.*

Trace evidence may either be deposited at a crime scene by the perpetrator or be carried away by him. He may leave toolmarks, bloodstains, hairs, fibers, and soil, for instance; he may carry away bloodstains, hair or fibers, glass fragments, soil, safe insulation, and other, similar traces on his person, clothing, or equipment. The investigator must keep these and other sources of trace evidence in mind and be diligent in his search for them at the scene, in the area, on the suspect, or on equipment used by the suspect.

The investigator should realize that in rare instances, usually because of a lack of sufficient amount of material, the laboratory is unable to render any opinion concerning the evidence. Negative findings of this nature can be avoided if the investigator follows the guidelines set forth in this chapter concerning the amount of a specific item of evidence to collect.

TOOLMARKS

The credibility and acceptability of toolmark evidence by the courts are reflected in the following excerpt:*

> Courts are no longer skeptical that by the aid of scientific appliances, the identity of a person may be established by fingerprints. There is no difference in principle in the utilization of the photomicrograph to determine that the tool that made an impression is the same instrument that made another impression. The edge of one blade differs as greatly as the lines of one human hand differ from lines of another. . . .

State of California v. Clark, 287 Pac. 18 (1930).

Definition

A *toolmark* is an impression, cut, scratch, gouge, or abrasion that is made when a tool is brought into contact with an object. A toolmark may be classified as a negative impression, as an abrasion or friction-type mark, or as a combination of the two.

Negative Impression

A negative impression is made when a tool is pressed against or into a receiving surface—for example, when a crowbar is used to pry open a door or a window. (See Figure 3.1.)

Abrasion or Friction Mark

An abrasion or friction mark is made when a tool cuts into or slides across a surface. This type of mark may be made by a pair of pliers, bolt cutter, knife, ax, saw, drill, plane, or a die that is used in manufacture of wire. (See Figure 3.2.)

Combination Mark

This type of mark is made when, for example, a crowbar is forcefully inserted into the space between a door and the door facing and

FIGURE 3.1   Negative impressions

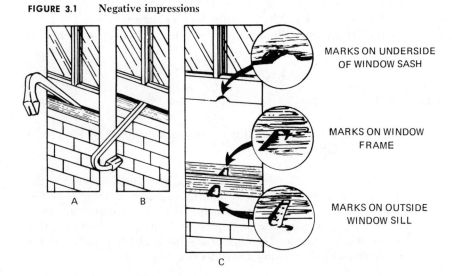

PRY BAR WORKED UP AND DOWN IN A AND B, LEAVING NEGATIVE IMPRSSSIONS AS SHOWN IN C.

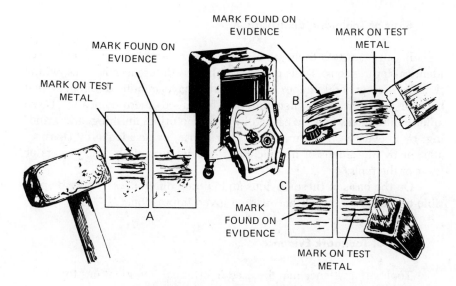

A. FROM GLANCING BLOW OF HAMMER
B. FROM PRY BAR PUSHED ACROSS SURFACE
C. FROM CHISEL DRIVEN INTO METAL

FIGURE 3.2 Abrasion or friction marks

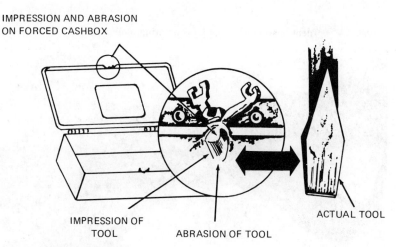

FIGURE 3.3 Combination mark

pressure is applied to the handle of the tool to force the door open. The forceful insertion of the crowbar produces an abrasion or friction mark, and the levering action produces a negative impression. (See Figure 3.3.)

50   Trace Evidence

Basis of Toolmark Identification

Because no two tools are alike in every detail, they will not leave identical impressions. Tools may have obvious differences in sizes, width, thickness, or general shape; but they also have minute differences that become apparent only when they are examined microscopically. These minute differences may be caused by manufacturing, finishing, and grinding; by uneven wear; by unusual use or abuse; by accidents; by sharpenings; or by alterations or modifications that were made by the owners or users of the tools.

On the basis of these obvious and minute differences, it may be possible to identify the tool that made a given impression.

Uses of Toolmark Evidence

Toolmark evidence may be used in criminal investigations to:

1. Link a person who utilizes a given tool with the crime scene, the commission of a crime, or some act material to a crime.

2. Establish whether a given tool or weapon found at a crime scene made a mark that is material to the crime. This knowledge is of value to the investigator whether or not the owner or possessor of

FIGURE 3.4   Identification of a tool by comparing toolmarks

FIGURE 3.5   Striation comparison

the tool is known, because it may eliminate the necessity of tracing a tool that, even though found at the crime scene, has no connection with the crime.

3. Establish a connection between similar items of evidence discovered in a series of crimes.

4. Determine whether a door or window was forced open from the inside or the outside.

5. Compare a toolmark from a crime scene with a toolmark found on the property, equipment, or vehicle of a suspect.

6. Facilitate and narrow the search for a given tool or weapon.

Examination Techniques

Protection of Evidence

The investigator arranges for police personnel to be posted to prevent damage to evidence until he has had time to examine it thoroughly. Doors, windows, transoms, and other openings with hinged or sliding

FIGURE 3.6
Ripped-open safe

FIGURE 3.7
Floor of safe ripped open

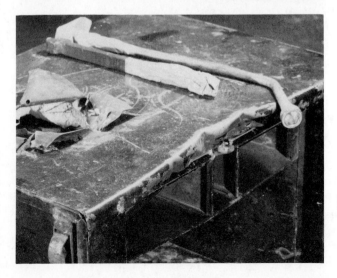

FIGURE 3.8
Tools left at the scene

doors or covers are not opened, closed, or handled in any way that is likely to destroy or mar minute toolmarks or fingerprints.

Searching for Toolmark Evidence

The investigator carefully examines every door, window, transom, skylight, and other opening that may have been used by a criminal as a means of entry or exit. Toolmarks are likely to be discovered at these points, particularly when forcible entry or exit has been made. The investigator pays particular attention to broken, forced, or cut locks, latches, and bolts, and the immediate area surrounding these fastenings. He systematically searches the entire crime scene and its vicinity for the tool that may have been used. He also examines safes, cabinets, desks, chairs, tables, or ladders for marks. The toolmarks are preserved even if no tools are found at the crime scene; the tools that made the marks may be discovered later. Toolmarks and tools are immediately and carefully noted and pertinent information recorded. They are included in any pictures and sketches of the crime scene.

Discovery of Evidence

Once discovered, a toolmark should be photographed as soon as possible. It should be examined visually to determine its gross appearance, thus providing the investigator with information concerning the type of tool or shape of tool he is to look for. The gross appearance of a tool impression may not be complete or well defined; for example, a hammer impression on a steel safe may not include the edges of the hammerhead, and so the shape of the head cannot be determined. In this case, all suspect tools that could have made the mark must be sent to the laboratory for comparison purposes.

If the surface bearing an impression of a toolmark has been painted, a careful examination may reveal that flakes of paint have been removed and may be adhering to the tool marking the impression. This might assist the investigator to eliminate a number of possible suspect tools. The pattern formed by the removal of the flakes of paint may also be of value. If a tool is found bearing paint similar to that of the painted surface, and the flake pattern appears to be identical in formation, the pattern should be photographed, since some of the flakes of paint might be loosened and accidentally removed from the tool while in transit to the laboratory. (See Figure 3.9.) The accurate matching of the pattern of paint on the tool with the pattern of the impression may be conclusive proof that the tool made the impression.

FIGURE 3.9   Matching paint chips from the scene and the suspect tool

The tool should never be fitted into the toolmark to see if it could have made the impression. Such a procedure may prevent the admittance in court of any evidence concerning the tool and its marks, or the paint on it and the object.

Processing the Evidence

A decision must be made as to whether the toolmark is to be removed for transmittal to the laboratory. Laboratory comparison is highly desirable, but wholesale removal of property or integral parts of valuable structures is neither desirable nor necessary. This is a judgment area, and the investigator's decision is based on the following factors:

1. Importance of the case
2. Importance of the toolmark in comparison with other available evidence
3. Distance of the crime scene from the police crime laboratory

Courses of Action

The investigator has three courses of action. He can remove the original evidence or the desired portion, cast and/or mold the evidence, or request that a technician from the police crime laboratory come to the scene and develop the evidence. In considering these courses of action, the investigator must also weigh the probative value of the original evidence.

Original evidence is more useful for scientific examination and evaluation and is less subject to attack in court than are reproductions. The investigator is often unable to make photographs and casts that represent the evidence sufficiently for identification purposes at the laboratory. In fact, some authorities recommended that casting or other methods of taking impressions of a toolmark should be used only as a last resort, since a casting can never be equal to the original impression. This is particularly so with regard to toolmarks made in soft materials such as wood, putty, or paint; many of the casting media most suited for these materials will not reproduce the fine details that are decisive for identification. Experiments have shown that scratches in paint caused by minute irregularities in the edges of a tool cannot be reproduced by an impression or a cast. However, if the original mark is compared with one made directly by the tool, the full proof against the criminal may be obtained.

Removal of Evidence

If a toolmark can be removed for transmittal to the laboratory, the investigator should take several precautions. A sufficiently large piece of the object should be removed to prevent damage to the toolmark through splintering, bending, twisting, or abrading. Any tools found at the crime scene will be transmitted to the laboratory along with the items of original evidence.

An item removed as evidence should be clearly marked with the case number, the investigator's initials, and the date and time of removal. The evidence is also marked to show the inside, outside, top, and bottom surfaces, and the area bearing the toolmark.

The investigator removes, for example, such evidence as the marked portion of a door, window sash, window sill, or door sill and that portion of the window or door frame that is adjacent to the marked area; and any window latch, door latch, bolt, hasp, or lock that has been cut, broken, or forced during entry. For private property, the investigator contacts the owner or custodian of the property and makes acceptable arrangements to return, replace, or pay for items that must be removed as

56 Trace Evidence

evidence. He makes certain that competent witnesses are present during the removal of the evidence in order to prevent later claims against the police department and to verify the original condition of the evidence.

If the surface bearing the toolmark is painted, samples of the paint should also be transmitted to the laboratory. In many cases, even though no paint can be seen adhering to the tool, enough minute particles may be recovered from it to permit an analysis and comparison at the laboratory.

It is sometimes possible for the investigator to furnish the laboratory technician with information concerning the angle at which the tool was held when it made the mark. If this determination is possible, information concerning the angle formed by the tool and the surface of the object, and the angle of deviation from the longitudinal direction of scrape marks, will materially assist the technician.

If a toolmark is on metal and cannot be removed, samples of the metal should be obtained and transmitted to the laboratory. Particles of metal may adhere to the tool in addition to paint, and may be analyzed and identified by the laboratory technician.

If pieces of wire that have been cut are to be sent for examination, the suspect end of the wire should be clearly marked. In obtaining the wire for laboratory examination, it must not be cut with the suspect tool. (See Figure 3.10.) Stolen articles such as an automobile radio, which can-

FIGURE 3.10 Strands within a piece of wire displaying toolmark impressions

not be positively identified by the owner, can often be identified as having been originally mounted in the automobile by matching the cut ends of the wire remaining attached to the automobile to those on the radio.

Photographing

The investigator carefully photographs a toolmark before it is moved, cast, or molded, or disturbed or altered in any way. Photographs provide a permanent pictorial record of the evidence in its original state and location, identify original evidence with any casts or molds that may be made, and satisfy legal requirements for records of original evidence.

The investigator makes two types of photographs:

1. A picture that shows the toolmarks and enough of the surface on which it is located to identify them positively
2. A closeup suitable for enlargement to show minute details of the toolmark

Initial photographs should show the mark as it actually appears and its overall relationship to other objects at the scene. Subsequently, the investigator should include an ordinary ruler and marking data in each picture to provide laboratory technicians with a scale of measurement for examination and comparison purposes.

Casting

When to use. The investigator makes a cast or mold from a toolmark only when he has a good reason for not removing the original evidence.

Materials. An impression found on wood or on a metallic surface may be cast by means of modeling clay or plasticine. These materials do not require any special preparation before use, nor are they likely to damage a toolmark if the first casting attempt is unsuccessful. A reproduction of the toolmark itself may be made from this cast, using plaster of paris.

Ordinarily, it is not necessary to reproduce a mark on a wooden surface because of the ease with which the original evidence may be removed. Such reproduction is undesirable, because a cast of a toolmark on a wooden surface does not usually disclose enough identifying characteristics to permit identification. Where it is necessary to make a cast of a mark, the material best suited for reproduction will necessarily be determined by the shape and type of mark to be reproduced. Flat toolmarks, hammer marks, chisel cuts, and pry marks may be reproduced using a variety of materials. Toolmarks in wood, where undercuts are present, will have to be reproduced with a flexible material. Suitable media, by brand or generic name, include among others the following: Kerr Permlastic; silicone rubber; Dow Corning Silicone Rubber "RC 900"; Kerr Perfection Impression Compound; plasticine; Castoflex; moulage and Posmoulage; Wood's metal; and plaster of paris.

Caution: The investigator should not attempt the casting or molding of a toolmark unless he has repeatedly practiced the particular method on a similar wooden or metallic surface of no evidentiary value. He should take sufficient care and time to ensure a usable reproduction. He should never release the surface bearing the toolmark until he has obtained an accurate reproduction and cleared the release with appropriate legal authorities.

Transmittal to the Laboratory

Each piece of evidence should be marked for identification and wrapped separately. The evidence samples should not share the same package unless all danger of intermingling has been eliminated. Toolmark evidence should be so wrapped and packaged that the toolmark and the tool will not be subjected to damage or loss of trace particles.

Laboratory Examination, Findings, and Value

The examination of toolmarks and the laboratory's findings are based on the same general principles and techniques as those used in fingerprint or firearm identification. Just as nature never repeats two things in an identical manner, things made by man can never be identical. In the laboratory, test marks are made with suspected tools on materials similar to those on which the toolmarks are present. Depending upon the circumstances, the test marks are compared with the suspect toolmarks by using the comparison microscope.

In many cases, the laboratory will find that the suspected tool made the toolmark found at a crime scene. However, it must be realized that such findings are not always possible. Sometimes the material on which the toolmark is found is such that it fails to record the minor imperfections needed to positively identify the tool as having made the evidence mark. In such cases, the examination may provide other information of value as investigative leads.

FIGURE 3.11   Wire found at scene compared to wire found on burglary suspect

An examination of toolmarks, even lacking a suspect tool, can be of value. A series of burglaries may be linked together by a comparison of the toolmarks found at each. A matching of the lengthwise markings on two pieces of wire may indicate that both were manufactured at the same time, having been drawn through the same die by the wiremarking machine. The suspect's possession of a piece of wire matching exactly the wire found at a crime scene would indicate that possession was more than accidental. (See Figure 3.11.) Wood shavings produced by a drill, plane, or other tool capable of producing wood chips may be identified with the tool producing them.

SERIAL NUMBERS

Serial numbers, consisting of numerals, letters, or symbols, individually or in combination, are placed on many manufactured articles, providing an easy means of distinguishing one from another. Purchasers of articles not serially numbered during manufacture often place their own private marks on them, for identification if they are stolen; but unless such special precautions have been taken, serial numbers may be the only way to establish ownership.

Serial numbers or marks may be stamped, molded, etched, or engraved. Many articles having component parts, such as automobiles, weapons, and watches, bear serial numbers on several of the parts. An investigator finding an object from which a serial number appears to have been removed should search it for other numbers, which are usually found in relatively inconspicuous places.

Investigative Leads

Individually owned items bearing serial numbers can usually be traced from the manufacturer through the wholesaler and jobber to the retailer, and from there to the original purchaser. Large industrial corporations, however, which buy large quantities of items, often do not initially record individual serial numbers, and shipments are accounted for by lot numbers, shipping and receiving documents, and sometimes other methods, to expedite the movement of supplies. Often, the manufacturer of a serially numbered item can furnish the lot number or other recorded information concerning items purchased by these corporations. As the bulk shipment is broken down for issue to units, the serial numbers are normally used on records and for identification. A persistent investigator, armed with lot numbers or shipping-document numbers, is usually able to narrow the search to the unit of ownership.

Restoration

There is no easy way to determine whether a serial number that has been removed can be restored, and there is no field method by which initial experimentation will reveal clues. All items from which serial numbers appear to have been removed should be sent to the laboratory for technical processing. A decision that a serial number cannot be restored is not valid in the absence of laboratory effort at restoration. Neither the material of which the article is made, be it wood, leather, metal, or other substance, nor the method used to apply the serial number, whether by stamping, molding, etching, or engraving, automatically precludes restoration. Serial numbers have been restored under the most adverse conditions; conversely, restoration attempts have been unsuccessful when conditions appeared to be most favorable. (See Figure 3.12.)

FIGURE 3.12 Number restoration

JEWELERS' MARKS

Although jewelers' marks are not serial numbers, investigative leads are available from them, and they can be used in tracing stolen property.

When an item is placed in the hands of a jeweler for repair, it is customary for him to place a small identifying mark in an inconspicuous place on it. The mark is usually inscribed with an extremely fine engraving tool under magnification, so that it is visible only under comparable magnification. Jewelers in the same geographical area are usually familiar with one another's markings. When the jeweler who inscribed the item has been located, he may be able to identify the person who brought the object to him. An investigator should not overlook the possibilities inherent in a jeweler's mark.

LAUNDRY AND DRY-CLEANING MARKS

All laundry marks, whether they are placed on clothing by a laundry, a dry cleaner, or an institutional laundry, are designed for one purpose: to identify the owner of the particular item of clothing. In crimes of violence, parts of clothing bearing laundry marks may be torn from the person of the suspect. In some crimes, such as holdups, outer clothing such as uniforms, overalls, or coveralls may be worn as a disguise and discarded as soon as possible after perpetration of the crime. Such clothing has been found hidden in trash bins and garbage cans, and the suspect has been identified through the tracing of laundry marks. An investigator may also make use of laundry marks to identify an unknown victim, or to establish ownership of property.

Examination

Any garment that comes into the possession of an investigator under circumstances that require the identification of the owner should be carefully searched for laundry marks. Special care must be exercised during the search if the garment is to be processed for other trace evidence, such as hairs, fibers, or soils, to ensure that such evidence is not dislodged and lost. All parts of the garment should be searched so as not to overlook a hidden mark that may have been placed in an unusual location on the garment. After an initial search, whether it is successful or not, another search should be made with strong crosslighting, to discover old, faded markings; and an additional search using ultraviolet light may reveal markings that are not otherwise visible.

The garment should be sent to the police crime laboratory for further examination, with information indicating any discovered markings and the means used to discover them. This will aid the laboratory technician in his examination and preclude duplication of effort. By the use of light sources of varying wavelength, the laboratory technician may discover markings that otherwise would remain invisible.

Tracing the Mark

The investigator must be prepared to spend long hours in the time-consuming task of going through the records of the laundry or dry cleaner, since such businesses do not have the personnel or the time for this search.

Commercial laundries and dry cleaners are often required to make their marks a matter of record in the local police department. However, if such records are not maintained, visits must be made to local laundry and dry-cleaning establishments. If local searches are unsuccessful, a request to the Federal Bureau of Investigation may be fruitful. The FBI maintains a complete filing system of all invisible laundry-marking systems. If their files do not contain information on the mark in question, they may be able to furnish information of value for further investigative activity.

BLOODSTAIN EVIDENCE

In crimes of violence, bloodstain evidence, if properly handled, is of great value to the investigator. Sometimes the evidence may be found in the form of fluid or clotted blood, but more often it is discovered in the form of fresh or dried bloodstains. Blood clots and bloodstains require careful examination, for none is so characteristic in appearance that the investigator can definitely ascertain its origin.

Bleeding

The body has a defense mechanism against excessive bleeding; as soon as bleeding starts in any great quantity, the blood pressure automatically drops, and consequently, the rate of bleeding slows. Upon death, the blood pressure falls to zero and the bleeding ceases. The dead do not bleed, except from a large wound located in such a position on the body that there is drainage due to gravity. This drainage, a mixture of blood and serum, and frequently other materials, is generally quite dark in color and may accumulate in considerable quantity. The amount of blood around a body may be important. A large amount of blood that has apparently issued from a small wound would indicate that the victim survived the attack for a considerable length of time.

Color and Visibility

Blood normally begins to clot after three to five minutes. As it dries, the clot darkens in color until, at the time it is completely dry, it becomes reddish-brown or dark brown. An old, dried blood clot may become so dark as to be almost black. Because of mold, putrefaction, or chemical changes, some bloodstains may appear to be black, green, blue, or grayish-white in color instead of the usual reddish-brown. The color of the blood should be noted.

If the blood falls on a porous material, such as cotton, wool, blotting paper, porous brick, or soft wood, the original color may be altered by the absorption of the blood into the substructure of the porous material.

A bloodstain on a dark background may be difficult to recognize, but use of a flashlight may reveal it even in daylight, for under artificial light a dried bloodstain may appear as a glossy or flat varnish, against a dull background. Indoors, where the amount of light is limited, dried bloodstains on a dark-colored floor may be made more visible by shining the flashlight beam parallel to the floor rather than perpendicular.

Shape, Persistency, and Age

The shape of the bloodstains may provide important information about the circumstances of a crime. The height from which a drop of blood fell may be determined, in many cases, from the appearance of the bloodstain. If the height of the fall is short, 6 to 12 inches, the bloodstains may appear as circular disks on a smooth surface. If the height of the fall is from 12 to 60 inches, the edges of the bloodstains may be jagged. This jaggedness increases in direct relation to the height; the greater the height, the more jagged the edges. If a drop falls from a considerable height, 2 or 3 yards, it may splash upon impact and form many small bloodstains, generally concentrated around a larger central stain, giving a sunburst appearance. Drops of blood that strike a surface at an angle may bounce or splash, leaving a large initial bludgeon-shaped blot with a series of smaller blots (similar to an exclamation point) trailing off in the direction of fall. Usually, the larger splash is made first and the smaller ones afterwards.

The investigator should not draw hasty conclusions concerning the direction of travel of the wounded person from the appearance of the bloodstains. The material upon which the bloodstain rests may alter the original shape of the drop as it strikes, and bodily movement may actually cause blood to fall in the direction opposite that of travel.

Occasionally, blood may be identified on a garment that has been washed, if the washing is not thorough. If, however, the laundering process is thorough, done with soap and hot water, residual bloodstain traces usually cannot be identified as blood. Washing of the hands may fail to remove all traces of blood, especially under the fingernails and around the cuticle at the base of the nails.

Bloodstains usually clot in from 10 to 20 minutes, and by examination of the clot, the investigator may be able to estimate the period of time that has elapsed since it was made. However, there are no criteria by which to judge the age of a bloodstain with any degree of certainty, since clotting time may be altered by many circumstances. For instance, it is more rapid on a rough than a smooth surface. Oily substances may not only increase the clotting time, but alter the appearance of the blood. Even the peculiarities of the blood of an individual may affect the clotting time.

A single drop of blood that falls on a dry surface, such as a table or wooden floor, will usually dry completely in about an hour at room temperature. Blood that has collected in a pool dries slowly, depending upon the size and depth of the pool formed, and the temperature and humidity to which it is subjected.

Source

Microscopic examination may sometimes disclose the origin of the blood because of the presence of foreign particles. Mucus or hairs from the nostrils may be found in blood from the nose. Semen and genital hairs may be found in blood resulting from rape. Certain cells from the vagina may be noted in blood from menstruation. Although the presence of foreign elements or particles may lead to conclusions as to the place or origin of the blood, their absence does not necessarily invalidate such conclusions.

Role of the Investigator

The investigator must be aware of what can or cannot be accomplished by expert examination of bloodstain evidence. He must realize that the value that may be derived from the examination of such evidence will depend almost entirely upon his using proper methods in collecting, identifying, preserving, and transmitting the specimen to the police crime laboratory. All types of blood evidence should be forwarded to the laboratory by the most expeditious means available, in order to prevent deterioration.

If the specimen is found to be human blood, a determination will be made as to what blood group it belongs to. The blood of every human being belongs to one of four blood groups, O, A, B, or AB. This grouping is based upon the presence or absence in the blood of specific substances, either singly or in combination. The blood group is not changed by the lapse of time or by disease. In the continental United States, approximately 43 percent of the population belongs to group O; 40 percent to group A; 11 percent to group B; and 6 percent to group AB. Medical laboratory examination may also determine blood-group subgroupings, RH factor, and the presence of some diseases.

Determination of the grouping of dried bloodstains is considerably more difficult than that of liquid blood. The age of the dried bloodstain or degree of exposure to direct sunlight, extreme temperatures, or other natural conditions may produce changes that reduce the possibility of successful grouping.

For blood-group testing, it is necessary to have more than a small spot of blood. The tests should be conducted on actual dried clots of blood, but if specimen quantity is limited, the grouping tests may have to be eliminated entirely and the examination limited to the chemical and precipitin tests. The greater the amount of the specimen furnished, the greater will be the possibility of obtaining maximum information during laboratory examination. A fairly heavy bloodstain measuring ½" by ¼" is generally sufficient for a conclusive grouping determination; but specimens that do not conform to this size must not be arbitrarily discarded by the investigator as unworthy of examination. And in the case of a putrefied or embalmed corpse, grouping tests may be made on the tissues.

If it is impossible to obtain a sufficient specimen of blood for grouping, it may be possible to obtain specimens of other body fluids containing the same substances that permit blood to be grouped. Between 65 and 80 percent of the population are secretors—people whose body fluids, such as saliva, perspiration, and semen, permit accurate typing as to blood group. Accurate groupings have been made of the saliva found on cigarette butts. The possibility of utilizing the dried remains of body fluids other than blood should never be overlooked.

Accurate blood grouping depends upon the ability of the examiner. The tests demand that the examiner have extensive practice and experience, a thorough knowledge of the necessary control tests and techniques, and the ability to recognize all reactions. A laboratory's failure to provide information concerning bloodstained evidence is usually the result of an investigator's failing to submit suitable specimens.

Marking of Evidence

Stained evidence found by the investigator should be immediately marked by him so that he can positively identify it at any subsequent time. If feasible, he should place his initials directly on the evidence, in an inconspicuous place, away from the evidence stains. The marking of initials directly on evidence is applicable not only to clothing, but also to metallic objects such as axes, knives, crowbars, and hammers. A tag should be attached to the evidence for further identification. If the evidence cannot be marked, the identifying data should be noted on the container in which it is placed. The investigator should record all details of the marking of evidence in his notes.

Handling of Evidence

Drying. A bloodstained article should not be packed for transmittal to the laboratory until it has dried thoroughly. The drying must be a natural process; that is, heat or electric fans should not be used. If heat is applied to a bloodstain, physical changes will take place within it that will interfere with its examination in the laboratory. The airstream from a fan may remove hairs, fibers, or other microscopic particles from the exhibit that might have a considerable bearing on the investigation, or it may blow extraneous material onto the bloodstain.

Bloodstained Articles. Clean wrapping paper should be used for packing bloodstained articles. Each should be wrapped individually before it is boxed for transmittal to the police crime laboratory, in order to prevent stains or other microscopic evidence from being transferred from one article to another. When bloodstained clothing is submitted to the laboratory, the entire garment, where possible, should be transmitted, in order to ensure a complete analysis. If the bloodstain appears on a large object, such as a rug or a drape, the size of which makes its transmittal impractical, the bloodstained area may be removed. However, for control testing, some of the unstained material from around the bloodstain should be included. In all cases, as much of the dried stain as possible should be submitted.

Samples from Fixed Objects. Where bloodstains appear on fixed objects, or objects too bulky to transmit to the laboratory, a different procedure should be used. That portion of the object bearing the stain should be cut off and sent to the laboratory, if possible. Stains on objects that cannot be cut, such as concrete floors or metal safes, are lightly swabbed with a sterile cotton swab that has been dampened with either saline or distilled water. This loosens the crusty portion of the stain so it can be collected on a clean piece of paper. The paper is then folded and placed in a vial or other suitable container. The remainder of the stain is taken up by rubbing it with the damp swab. The swab is allowed to dry, then placed in a suitable container. Do not scrape a bloodstain off a surface until the methods above are tried, because the sample may fragment and harden rapidly, making certain tests at the laboratory difficult or impossible to perform.

Porous Articles. If a bloodstain appears on a porous article, such as wood or earth, the stain and a portion of the material upon which it appears should be removed, so that proper control tests may be made. The material removed should be placed in a clean pillbox or similar container, properly labeled on the outside with identifying data and sealed to prevent loss through leakage.

Liquid Blood Samples. It is often desirable to obtain liquid blood samples from people involved in an incident and forward the samples to the laboratory with other evidence. Blood samples should be drawn by a physician or a trained medical technician. Precautions must be taken to prevent contamination of the samples. Hospitals, police dispensaries, and the like have available sterile containers that may be used in transmitting blood samples to the laboratory.

The quantity of liquid blood required for laboratory examination is approximately five cubic centimeters, or about ⅙ of an ounce. No preservatives should be added to whole blood, as these interfere with subsequent blood tests. If a delay is anticipated between the time the blood is drawn and the time that it can be forwarded to the laboratory for analysis, it should be refrigerated; however, freezing must be avoided.

A death by drowning in fresh or salt water may be determined by a chemical examination of blood. To make such a determination, the laboratory must have blood specimens from both the left and right sides of the heart. Approximately five cubic centimeters is sufficient for each specimen. The services of a medical laboratory will be required to obtain such specimens and make the determination.

OTHER BODY FLUIDS

Other body fluids, in addition to blood, are often found at the scenes of crimes of violence. The other body fluids include semen, saliva, urine, perspiration, pus, milk, nasal mucus, and tears. Crime laboratory attention is primarily concerned with semen and saliva.

Like other evidentiary trace materials, all these bodily fluids must be examined by technicians in the laboratory. If they are to be of any value, their initial handling by the investigator is most important. A laboratory's inability to provide information concerning body fluids can usually be attributed to insufficiency of the sample and a failure to supply adequate standards and controls.

The same factors that make it possible to distinguish one person's blood from another's are present in the cells of every organ of the body and in almost every body fluid. Medical researchers originally demonstrated the presence of the grouping factors by examining known wet specimens under ideal laboratory conditions. They discovered, for example, that the concentration of the grouping factors in saliva and semen secretions was relatively high and that the concentration in tears, urine, and perspiration was very low. Following the medical recognition of the results of this research and the standardization of the identification techniques, the identification procedures were applied to the stains of such fluids when encountered in medicolegal cases.

A secretor is a person whose body fluids, as well as blood, may be grouped; a nonsecretor is one whose blood may be grouped, but whose body fluids may not. Researchers established that the proportions of secretors and nonsecretors are approximately 65–80 percent to 35–20 percent respectively. The saliva is the most suitable material for distinguishing secretors from nonsecretors. It is easy to obtain, and if the person is a secretor, the relatively high concentration of group specific substances will be easily noted. The ability or inability to secrete group specific substances is a constant function of the organism, so if they are present in the person's saliva, they will be present in almost every other fluid of his body.

In criminal cases, body fluids usually appear as stains on clothing, bedding, upholstery, or similar objects. In examining a body-fluid stain, the techniques and methods of identification are the same as for bloodstains, but identification is more difficult depending on the nature of the stain, its age, and the presence of interfering factors. When handling evidence on which stains may have been caused by body fluids, the investigator will exercise the same precautions prescribed in handling bloodstains.

Importance

The relation between blood and body fluids offers many avenues of approach to the investigation of an offense. For example, where a blood sample is not available because of putrefaction or for other reasons, tests may be made on other body fluids. A cigarette butt found at the scene of a crime may contain dried saliva that can be grouped and compared with the blood group of a suspect.

Although the actual examination of body-fluid evidence is left to a qualified crime laboratory technician, the investigator should ensure that such evidence is transmitted properly and expeditiously to the laboratory. He should also be aware of what can and what cannot be expected of a laboratory examination of this type of evidence.

Identification

The group substances in liquid or dried saliva and some other body fluids may be classified into groups A, B, AB, and O.

As in the case of a bloodstain, the identification of a body-fluid stain may be negative, or an inconclusive opinion may be reached. Nevertheless, the possibility that valuable investigative leads may result from the expert examination of a body-fluid stain should not be overlooked. For example, if a dried saliva stain is found that contains group substances, it could not have been derived from a person whose saliva does not contain these substances. However, if the dried stain is free of these substances, two possibilities exist: either the saliva came from a person who is a nonsecretor, or the specimen is free of the grouping substances because of contamination or deterioration, and a definite opinion is not possible.

Some men's semen can be typed, where the specific factors of the blood are carried in the fluid. Saliva normally provides a reliable typing. The typing of urine and other bodily fluids is subject to many more variables, and consequently, reliable typing is more difficult.

Seminal Fluid

In case of rape or sexual assault, where it is alleged that the attacker had an emission, the identification of the seminal fluid is of paramount importance. Semen is a colorless, tenacious fluid, produced in the male reproductive organs, which is usually discovered in the form of stains on clothing, bedding, or similar articles. Fresh, undried seminal fluid has a characteristic odor and contains thousands of minute organisms known as spermatozoa, which die as the semen dries, although they preserve their shape indefinitely if they are not destroyed by handling. In its dried state, semen appears as a grayish-white, sometimes yellowish stain and imparts a starchy stiffness to the portion of the fabric on which it is contained.

Specific tests for semen involve the identification of the spermatozoa and positive chemical testing of the stain; therefore, articles that are suspected of bearing seminal stains should be handled so that there is no friction whatever against the stains. For transmission to the criminal investigation laboratory, the articles may be placed between sheets of cardboard or similar material, and secured in such a manner that friction is avoided. Under no circumstances should a stained area be rolled or folded. Stains should be allowed to dry.

The inspection of evidence under ultraviolet light is sometimes useful in revealing the location of seminal stains, because of their fluorescent qualities. Laundering almost always removes traces of seminal stains.

Saliva samples should be obtained from suspects and victims involved to determine if they are secretors, and if so, what their blood types are. This is done by having each person chew on a separate piece of thin gauze about 2" by 4" for about five minutes at the back of the mouth. The gauze is then allowed to air-dry completely, placed in a sterile container, and sent to the laboratory. If the gauze is not dried, enzymes in the saliva will destroy the blood-group substances, which may lead to an incorrect result in subsequent testing.

Other Stains

Objects suspected of bearing stains from body fluids other than semen should be handled with the same precautions and transmitted to the police crime laboratory in the same manner as bloodstained articles.

HAIRS AND FIBERS

The value of hairs and fibers as evidence in criminal cases has been clearly recognized. Hairs and fibers are seldom conclusive as evidence, but in conjunction with other details, have proved to be essential aids to the investigator. The investigator must capitalize on the importance of this type of evidence during the initial phase of the investigation.

Importance

The origin and texture of hairs and fibers found at the crime scene, or upon the body, clothing, or headgear of a suspect or victim, may be exceedingly important as evidence, particularly in homicides and sex crimes. Hairs may be pulled out during the crime and found at the scene or on the victim. Hair and fiber transfer may occur during any physical contact between the suspect and the victim. Hair may fall out under conditions that the suspect is not aware of and unable to guard against. Properly handled by the investigator, hair and fibers may provide excellent investigative leads and add to the evidentiary facts being assembled.

Hair

Structurally, a hair is composed of the cuticle, cortex, medulla, bulb or root, and tip, each of which provides the laboratory examiner with definite information. The cuticle is the outer surface of the hair. When seen under the microscope, it appears composed of scalelike flakes, each overlapping the next, similar to the overlapping of shingles on a roof or scales on a fish. These flakes are known as cuticular scales. The cortex is the inner portion of the hair and contains the pigmentation or coloring. The medulla may be described as the core or center portion of the hair shaft and, under magnification, appears as irregular, spinal, chain, or a continuous dark line of varying width running up the center of the shaft. The examination of any or all three of these sections may reveal any one of a number of personal characteristics of its source. Other important information may be gained through visual, microscopic, and chemical examinations of the root or bulb, the shaft, and the tip of the hair. (See Figure 3.13.)

Hair Examination

Through examination, the police crime laboratory will normally determine first if the hair samples are of animal or human origin. If the

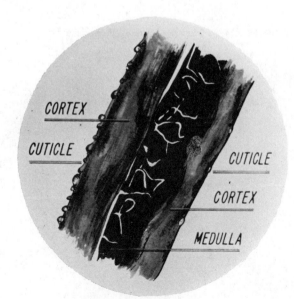

FIGURE 3.13
Shaft of hair; note foreign particles on cuticle

origin is an animal, a general determination as to species may be made.

In the case of human hairs, laboratory determinations may include:

1. The race of the person the hair originated from—Negroid, Mongolian, or Caucasian
2. The area of the body surface that the hair originated from—head, face, chest, axilar, limb, or pubic
3. How the hair was removed—naturally or forcibly
4. How the hair had been treated—bleached, dyed, waved
5. Whether the hair was cut with a dull or a sharp instrument, how recently it was cut, or if it was crushed or burned

Usually, examination will not permit conclusive determination of age or sex.

Laboratory hair-comparison conclusions will generally be stated in one of three forms: (1) that the hairs are dissimilar and did not originate from the same person; (2) that the hairs match in terms of microscopic characteristics, and originated either from the same person or another whose hair exhibits the same microscopic characteristics; or (3) that no conclusion could be reached. (See Figure 3.14.)

Fibers

Contact between two pieces of fabric can seldom be made without an interchange of fiber material. In cases involving physical contact, both

FIGURE 3.14   Microscopic hair comparison

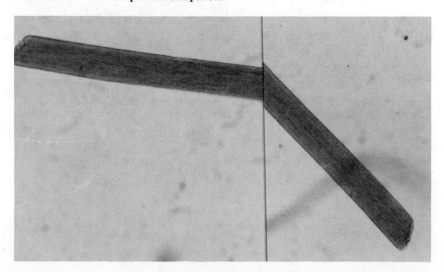

the victim's and the suspect's clothing may intermingle and exchange fibers. In burglary cases, contacts of clothing with objects should also be considered in the examination of the crime scene and the suspect's clothing for fiber evidence. Points of entry such as windows, or means of access to roofs, ladders, or drain pipes, may reveal fiber traces of value to the investigation. Clothing fibers and fibers from the upholstery of automobiles may also be transferred from one to the other.

Fibers are classified as being mineral (glass, asbestos), vegetable (cotton, linen, hemp, jute), synthetic (rayon, nylon, Orlon, Dacron), or animal (wool, silk, furs of all types). Classifications vary greatly as to color and type of processing, yarn and thread composition, and end use. Fabrics, tapes, ropes, and similar end products may be subjected to fiber examination.

Fiber Examination

Fabric that is composed of knitted or woven yarns (fibers twisted together) will be grossly examined in terms of color, composition, and construction. Questioned fabrics may be determined to be similar to known fabrics. Positive identification may be made where a questioned piece of fabric may be fitted back into the known fabric, by matching broken ends of yarn together.

Fibers will be identified as to type, color, and matching characteristics based on laboratory microscopic, microchemical, and melting-point examinations. Generally, fiber matches are not positive evidence, and require substantiation with other corroborative evidence.

Tape examination, similar to fabric examination, generally involves the matching of the ends of pieces of tape used at the scene of a crime with the end of tape on a roll found in the possession of a suspect.

Cordage, in the form of ropes and strings, is examined in terms of composition, color, diameter, and construction. The known is compared with the unknown, and occasionally ends may be matched or the manufacturer determined.

Handling and Transmittal to the Laboratory

The most difficult task facing the investigator is to initially locate hair and fiber evidence at the scene of the crime. The search must be thorough, detailed, and exacting. Obvious locations to search include headgear and clothing, with particular attention being given to linings, pockets, and cuffs. In addition to the general crime scene area, other areas to search may include the victim's body (particularly in sex crimes), underneath the fingernails, and any upholstered surfaces.

Of paramount importance is the basic rule, *avoid contamination of evidence*. Hair and fiber evidence is particularly susceptible to cross-contamination, and the investigator must ensure that evidence gathered from the suspect and from the victim is not intermingled. It must be individually collected, properly marked, placed on a clean piece of paper, which is folded and put in a clean container, and properly separated during packing for transmittal.

Detailed examination for hair and fibers should be left to the laboratory. When collecting known samples from the victim or suspect, gather a sufficient quantity. Twenty hairs or fabric strands are considered the minimum. Only a doctor should collect sample hairs from the body of a victim or subject. These samples should be obtained from the various parts of the body possibly involved in the crime.

Envelopes should not be used as containers, as hairs and fibers placed in them can easily sift out and become lost. The use of tape or glue to hold the hairs or fibers will interfere with subsequent laboratory examination.

FINGERNAIL SCRAPINGS

Fingernail scrapings are rarely exploited to the fullest advantage, although they may provide significant indications in some investigations. The causative agent of abrasions and scratches found on many parts of the body is frequently a fingernail. The face, neck, arms, thighs, and female genitals are the areas commonly attacked and should be subjected to careful medical examination. The form, extent, and location of abrasions will depend on the circumstances in each case.

Sexual Assaults

Resistance offered by the victim during a sexual assault often results in the gouging of skin by the assailant's fingernails. Consequently, important evidence can often be obtained by careful examination of the fingernail scrapings. If your suspect is the guilty party, minute particles of skin, blood, hair, and cosmetics may be found and identified as having come from the victim. Small fibers that can be identified as having come from the victim's clothing may also be found under the fingernails.

Occupation

Examination of the fingernails of a deceased person may aid in determining his occupation. Fingernails that are trimmed but not regularly manicured, and that bear scratches, may indicate some manual labor. Fingernails beveled, brittle, growing tight at corners, rounded at ends, and regularly manicured may indicate no manual labor.

Other Determinations

When a person is found hanging, all scratches on the head and neck should be critically examined to determine whether death is attributable to suspension or to the commission of a crime. Scratches must be carefully considered in relation to whether or not they could have been produced by the ligature. Scratches on the neck or face in the presence of a soft, pliable ligature should evoke suspicion, although it must not be forgotten that the dying person's vain attempts to loosen the ligature may have produced scratches with his fingernails. Consequently, important information may be obtained from fingernail scrapings of the victim.

Fingernail scrapings may also assist in identifying people who use narcotics and marihuana, and those suspected of poisoning their victims.

Procedures

In taking fingernail scrapings from a suspect or victim, never use a knife, file, or any other hard, sharp instrument. These are apt to cause bleeding, with resultant contamination of the nail scrapings. The best instrument to use is the blunt end of a wooden toothpick, using a different toothpick for each finger. When the scrapings are taken from one finger, the toothpick and scrapings should be placed on a clean piece of paper, which is folded up and packaged in a suitable container. Each container

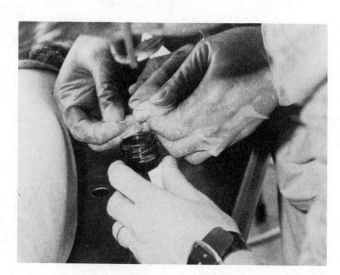

FIGURE 3.15
Fingernail scrapings from deceased person

should be marked to indicate the finger from which the scrapings were taken, and then sent to the police crime laboratory for scientific analysis.

SOILS, ROCKS, AND MINERALS

Soils, rocks, and other minerals may be found on the suspect, his shoes, his clothing, or his vehicle; on tools used in the crime; at the scene of the crime; and on the victim. These materials may provide valuable circumstantial evidence. Such evidence is often overlooked because investigators are unaware of its potential value or of the laboratory services available to exploit it.

Difference in Soils and Rocks

Soils and rocks vary in different locations throughout the world, and even within small local areas. The differences between two types of soil, such as sand and clay, may be readily recognized. The detailed differences and similarities between samples of similar soils or rocks, however, can be detected only by qualified laboratory technicians.

Cause of Differences

Over long periods of time, natural forces have produced variations in the composition of soil and rock deposits, both on and below the surface of the earth.

In addition, man has caused numerous soil and rock variations. In mining, agricultural, and industrial areas, there is an almost constant mixing and moving of soils and rocks. Materials and minerals are added to and taken from the soil. Crops may add identifiable particles to soils in which they grow. Soils and rocks are added to or taken from surfaces to make them level, thereby disturbing local geological patterns and further adding to the great variety of soil and rock differences.

Local variations may be caused by the addition to the soil of plant wastes, plant products, fertilizers, human and animal wastes, soil-conditioning materials, and many other substances.

Soils and Rocks as Evidence

One of the primary uses of soils and rocks as evidence is the comparison of samples from the crime scene with samples on the suspect's clothing or other possessions, to determine whether the suspect could have been at the crime scene.

Soil and rock evidence is more apt to be found when the offense was committed out of doors, or when the perpetrator walked or drove a vehicle on unpaved ground. Such evidence may be derived in the following ways.

The offender may leave at the crime scene small amounts of rocks and soils that he has carried there. In addition to linking him with the crime, these particles may provide clues to the offender's whereabouts and his occupation, and they may indicate whether he walked or rode to the scene. Also, the offender may pick up soil and rock materials at the scene, or he may both pick up and leave soil and rock evidence.

In a hit-and-run accident, the offending driver or his vehicle may pick up and leave incriminating evidence. The impact of the collision may dislodge mud, dirt, and accumulated debris from the undercarriage of the hit-and-run vehicle. The particles may be found on the road or on the body or clothing of a pedestrian struck by the vehicle. Or the vehicle may run off the road or along the shoulder of the road and pick up mud and dirt on its tires and undercarriage. Finally, the offending driver may dismount to survey the damages, to disengage his vehicle, or to make minor adjustments to it so that he can flee the scene, and in so doing, he may get mud or debris on his shoes or clothing or into his vehicle.

Other Mineral Evidence

Plaster and Building Materials

When a building has been broken into, a variety of building materials may be passed through or damaged. Materials may include plaster, plasterboard, insulation, sheeting, cinder block, mortar, and brick. Variations in composition, texture, and color, combined with the definite probabilities of transfer to the person or clothing of the perpetrator, make the collection and examination of these mineral materials essential.

Safe Insulation

The penetration of the walls of a safe may cause the insulation to be broken. If dust from the insulation is scattered about the scene, the perpetrator may get it on his clothing. Close examination of the scene may also reveal clearly defined footprints in the dust that has settled on the floor.

The substances utilized for insulating or fireproofing safes vary according to manufacturing specifications. The exact composition is a trade secret; however, the criminal investigation laboratory has data on the types used by major manufacturers and can conduct comparison and study of evidentiary specimens.

Dust

Dust from a house or other building may get on the shoes or clothing of a person who burglarizes or otherwise illegally enters it. Such dust may contain ingredients in proportions and combinations that will enable laboratory technicians to determine that it could have come from a certain place. For example, house dust may contain lint from clothing, bedding, drapes, and upholstery on furniture; wood fibers from the floor and furniture; paint and plaster particles; human and animal hairs and skin cells; soot; plant wastes; and the residue of floor waxes and cleaning compounds. Dust or grime from a vehicle repair shop is apt to contain metal particles and dust; dust from road surfaces; rubber and paint particles; lint and fibers from automobile upholstery and seat covers; and traces of lubricants, fuels, battery acids, and antifreeze compounds.

Collection of Evidence

Clothing, shoes, and other personal belongings that on detailed visual examination appear to contain soil, rock particles, grime, dust, mud, or similar substances should be secured as evidence. They should be individually wrapped; clean plastic bags are excellent containers for this purpose. The investigator should not attempt to remove evidence from shoes and garments, but should submit them to the police crime laboratory.

Tools suspected of having been used in the crime, or of having been at the crime scene, should be individually wrapped. Each should be placed in a wooden or heavy cardboard carton. The cutting or prying edge of a tool should not come in contact with any hard or abrasive surface. The wrapping must be secure enough to prevent the loss or contamination of foreign materials present on the tool.

If a foreign substance is found on an object or structure too large for shipment to the laboratory, the investigator should scrape the substance into a pillbox or similar container, which should be carefully sealed and marked for identification. Photographs, sketches, and notes should indicate the exact places from which such evidence was removed. Evidence from different areas must be kept separate.

The residue from under a suspect's fingernails may contain traces of substances from the scene or from the body or clothing of the victim. Scrapings should be taken from all the suspect's fingers, preferably before he has a chance to bathe or clean his nails. The scrapings should be kept separate in suitable containers, such as small pillboxes.

Collection of Comparison Samples

The investigator should always obtain samples of rocks and soils from the scene, to be used as standards for comparison with similar substances that may be from suspects. In serious cases, even if there is no suspect, he should always secure comparison samples. A competent witness should be present when they are removed, packaged, and marked.

Each sample should consist of about two tablespoonfuls of soil. In taking a sample, it is seldom necessary to go deeper than ½ to ¾ of an inch. However, if a footprint, tire track, or other indentation in question penetrates into subsoil that is different from the topsoil, it may be necessary to secure samples of both the topsoil and the subsoil.

If soil and rock evidence has been found on a suspect's shoes, a comparison sample should be taken from that portion of a footprint at the scene corresponding to the part of the shoe on which the evidence was found. Comparison samples should be taken from a footprint, tire track, or other evidentiary indentation only after a plaster cast has been made.

Comparison samples should be taken beginning at a starting point, which may be a footprint, a tire track, or a place where the suspect would probably have gotten soil on his clothing, shoes, vehicle, or tools. Samples should then be taken at varying distances from the starting point, to ensure that they are representative of the area's soil characteristics. A detailed sketch and notes pertaining to the samples taken, coupled with careful handling to preclude mixing of samples, will help to ensure subsequent evidentiary usefulness.

At the scene of a hit-and-run accident, the investigator should secure samples of the soil wherever the hit-and-run vehicle ran off the road. The road surface should be examined carefully for mud flakes that may have dislodged from the vehicle. If other vehicles are involved, samples for comparison purposes should be taken from the undercarriages of all of them. When the investigator finds a vehicle that he suspects of being the hit-and-run vehicle, he should secure soil and dust samples from its tires, undercarriage, brake pedals, and floor mats and submit them to the laboratory for analysis and comparison with other evidence and samples pertaining to the case. Comparison samples from road surfaces and shoulders and from vehicles involved in the hit-and-run accident should consist of approximately two tablespoonfuls of soil or dust from each place sampled. All the evidentiary materials gathered from the road should be collected and submitted to the laboratory.

The investigator should include in his sketch of the crime scene and in his notes the points from which soil and rock evidence and comparison samples were taken. The sketch should contain both compass directions and measured distances.

Packaging and Handling Evidence

Evidence and comparison samples to be submitted to the police crime laboratory should be prepared in the following manner:

1. During the collection and packaging of evidence and samples, the investigator should be careful not to mix any of the items of evidence with each other or with the comparison samples.

2. New or unused medical pillboxes, or small cardboard containers with tight-fitting covers (similar to those used for packaging ice cream), sealed tightly with cellophane or other adhesive tape, are suitable for packaging dry soil samples. If the soil samples are wet, they should be permitted to dry naturally in a dust-free room, not by artificial heat.

3. Soil and rock evidence or comparison samples that are believed to contain petroleum or other volatile materials should be packed promptly into small mason-type jars with tight-fitting lids. The lids should be screwed on tightly and the edges sealed with wax.

4. The individual containers, marked for identification and sealed to prevent leakage or contamination, should be packed into a large, substantial container for shipment to the laboratory.

5. Excelsior, cotton, or crumpled paper should be used to fill any vacant spaces in the package, to provide additional protection for the individual samples.

Laboratory Examination

Given adequate samples that have been correctly collected and identified by the investigator, laboratory personnel may either ascertain that two samples of rock or soil *could* have come from the same location (a more positive statement cannot be made, since it is possible, even though not likely, that samples from two different places may be identical), or determine that two specimens *could not* have come from the same location.

The laboratory cannot specifically determine where a sample of a questioned substance came from merely on the basis of known geological patterns. It may, however, provide the investigator with knowledge of the exact contents of a sample. This information may provide clues to local areas that have the same general type of soil or rock, as indicated by the laboratory report.

MISCELLANEOUS LABORATORY EXAMINATIONS

In the laboratory, trace evidence is subjected to a variety of examinations, some of which require the use of special equipment. The comments that follow are to familiarize the investigator with the general nature of these examinations and with the instruments utilized.

Ultraviolet Light

Ultraviolet light, or "black light," employs invisible radiation of a slightly shorter wavelength than normal, visible light. Ultraviolet light striking a surface is absorbed by some substances and, in turn, is radiated in a different-colored light. In a darkened room, the rays are initially invisible, but the effect on the substance is visible in that the light-emission phenomenon resulting is one of fluorescence. Numerous field uses may be made of ultraviolet light, and laboratory examinations may include the following:

1. The use of reflected ultraviolet light, or the fluorescence, can quickly and easily indicate similarity or differences in a variety of substances; for instance; the grayish-white fluorescence of semen may be differentiated from the yellowish-green of urine. And stains on clothing that are not visible when viewed by means of ordinary light may become visible when viewed under ultraviolet.

2. Glass samples that are similar in appearance and color under visible light may be differentiated under ultraviolet. During the manufacturing process, trace materials may be introduced into the constituents purposely or accidentally, producing different fluorescent or reflective qualities.

3. Cosmetics, such as fingernail polishes, lipsticks, and rouge, usually have distinguishing colors of fluorescence under ultraviolet light.

4. The paper and inks of documents show clearly their similarities or differences under ultraviolet. Glues and other adhesives used to reseal envelopes can usually be distinguished by an examination under ultraviolet light.

Infrared Light

Infrared light, also invisible radiation, is of a slightly longer wavelength than normal, visible light. It has no fluorescent effect and cannot be seen by the unaided eye, but rather requires examination through the medium of infrared viewing equipment. Some materials that do not show a color under visible light do absorb and reflect infrared radiation that can be detected photographically. Laboratory applications include the following:

1. Differences shown by infrared-light examination of paints and dyes often yield valuable investigative clues. Differences in opacity, transmittance, or reflectance are usually readily apparent.

2. Gunpowder residues on clothing may become visible even when obscured by dyes or stains, such as bloodstains. A photograph of the powder-residue pattern may permit proximity testing in the laboratory to determine the approximate distance that the muzzle of the firearm was held from the clothing at the time of firing. (See Figures 3.16 and 3.17.)

FIGURE 3.16 Normal panchromatic photograph of gunshot hole

FIGURE 3.17 Infrared photograph, showing powder residue around gunshot hole

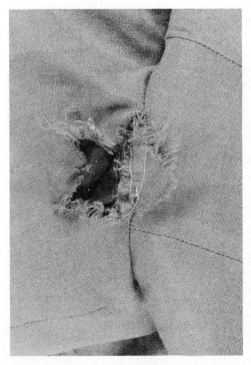
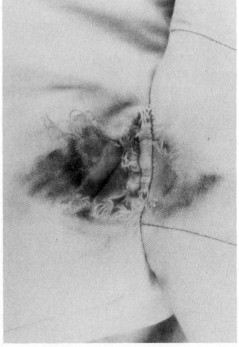

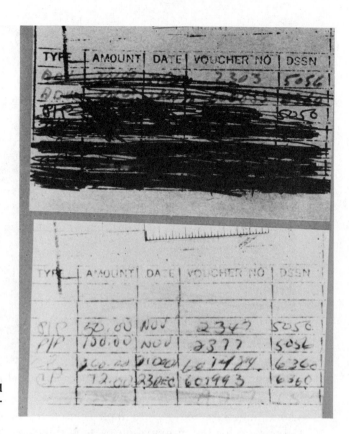

FIGURE 3.18
Overwriting removed through infrared photography

3. Inks also may be differentiated by examination under infrared illumination. Erasures on documents, as well as the writing or printing on charred documents, have been deciphered by using infrared light. (See Figure 3.18.)

Spectrograph

Minute quantities of evidentiary material are often analyzed utilizing the spectrograph, a laboratory instrument that produces a graph showing the basic constituents and trace elements of the substance examined.

When a substance is burned, it sends out energy, or light waves, which are observed as colors. In the spectrograph, this light is passed through a prism, and the resultant color pattern is focused on a photographic plate as a series of parallel lines of varying density. The different lines represent the different elements present, and their density or thickness corresponds to the quantity of the elements present. Analysis of this graphic portrayal will determine the nature of the substance and permit graphic comparison of two samples. Useful in analyzing primarily inor-

FIGURE 3.19   Emission spectrograph, for identification of metallic elements

ganic materials, the spectrograph is used in the criminal investigation laboratory for the examination of such things as paint, glass, dust, safe insulation, soil, wire, and other metals. (See Figure 3.19.)

Spectrophotometers

The infrared and ultraviolet spectrophotometers permit qualitative and quantitative analysis of a substance through measurement of its absorption of light rays of varying intensity. The laboratory application of this instrument lies primarily in the identification and analysis of substances such as drugs, dyes, inks, plastics, oils, rubber, and stains, and in determining color and making color comparisons.

Gas-Liquid Chromatography

Both qualitative and quantitative analysis of a substance, or of mixtures of several substances, may be performed on the gas-liquid chromatograph (glc). The material to be analyzed is injected into the glc, where it passes through a column that separates its components. As each component passes through a detector, the characteristic time to reach the detector and the relative amount of the component are recorded on a graph. Comparison of this graph with the graph of a known sample makes it possible to determine whether two substances are of the same or different compositions. This method is used primarily in comparison of paint fragments and of accelerants found in arsons.

4

Photography

INTRODUCTION

Photography properly performed can be one of the investigator's most valuable aids. To obtain maximum proficiency in the preparation of photographs, the investigator must be qualified to operate, or supervise the operation of, available photographic equipment.

CAMERAS

Types Most Frequently Used

Speed Graphic

This camera contains associated items adequate for most general purposes. The Speed Graphic, with its large 4″-by-5″ negative, its lens-coupled range finder, and flash synchronization, is adaptable to varying conditions and situations. The camera is usually equipped with an f/4.5 anastigmatic 127-mm lens and with a focal-plane shutter, as well as a between-the-lens shutter. The Polaroid Land 4″-by-5″ film holder fits the Speed Graphic and most other press-type cameras. With this equipment, the photographer can obtain 4″-by-5″ black-and-white or color photographs in one minute or less.

35 mm

The 35-mm camera is a small, compact, lightweight camera with a great depth of field, and it is simple to operate. The small size of the film requires the use of fine-grain materials and developers and exceptionally clean equipment and facilities. It is especially useful in situations that require fast operation. When used with appropriate copying equipment, it can be focused at close ranges to photograph small articles such as cartridge cases and bullets. The short-focal-length lens produces images with a greater depth of field than those lenses provided with the larger cameras.

Fingerprint

A fingerprint camera is a fixed-focus copying camera, equipped with a built-in light operated by photoflash batteries. It is portable and simple to operate. Images are recorded on the negative in actual size. Its uses other than as a fingerprint camera are numerous. For instance, it can photograph small documents or portions of large documents easily and quickly.

Motion Picture

One of the most widely used motion-picture cameras is the Bell & Howell. It is a 16-mm, silent, single-lens, magazine-load camera, spring-driven and able to take color or black-and-white film. The normal lens is a 25-mm f/1.9 anastigmat, which focuses from one foot to infinity. Accessory lenses are a 17½-mm (wide-angle) f/2.5, which focuses from 2 feet to infinity; a 51-mm (long-focus) f/3.5, which focuses from 2½ feet to infinity; and a 102-mm (telephoto) f/4.5, which focuses from 4½ feet to infinity.

Lens Speed

The speed of a lens is the maximum amount of light that the lens will transmit. This is calculated by using the inverse-square law and the ratio of the diameter of the lens to its focal length.

The *inverse-square law* states that the intensity of light diminishes inversely with the square of the distance from its source. If an object is placed one foot from a light source, a certain amount of light will strike it. If the object is moved so that it is two feet from the same light source, the intensity of the light striking it is only one-quarter as bright. If the object is moved four feet from the source, the intensity is only one-sixteenth the intensity at one foot.

When light passes through a lens, it forms an image. The area in which this image is formed is called the focal plane. The focal length (f) of a lens is the distance from the optical center of the lens to the focal plane, when the lens is focused on infinity.

To find the speed of a lens, divide the diameter of the lens into the focal length. For example, if a lens has a diameter of 2 inches and a focal length of 8 inches, then the speed of the lens is 4 (8 divided by 2). This is expressed as an f-number, indicated in this case by the symbol f/4.

Aperture and Shutter

The diameter of a lens can be reduced by including a diaphragm in the lens system. This reduces the usable diameter and consequently the amount of light transmitted by the lens. In the foregoing example, if a diaphragm reduces the aperture diameter of the lens to 1 inch, its speed would now be f/8—the focal length (8) divided by the diameter (1). Thus, in consonance with the inverse-square law, only one-quarter the amount of light is admitted.

The iris diaphragm, in use with the better lenses, consists of a number of overlapping leaves of the thin metal, one end of each leaf being fixed and the other fastened to a revolving ring, so that as the ring is turned, it will open or close the leaves, thus varying the size of the opening. Normally, the diaphragm opening is marked with the f-system.

The relative exposure with any two adjacent f-numbers for the same lighting conditions is 1 to 2. The larger number admits one-half the light and requires twice as long an exposure. For example, f/5.6 admits twice as much light as f/8, and the exposure would be half as long. As the stop openings become larger, the f-numbers become smaller.

The shutter serves to vary the time interval during which light is permitted to enter the lens. The time interval required for a good picture depends primarily on two factors: the lighting conditions, and the amount of motion to be arrested. When a shutter is released, it will open for a definite period of time and then close. The length of time the shutter is open is called the shutter speed, and is usually a fraction of a second. The faster the movement of an object being photographed, the faster the shutter speed will have to be to record an image free from blur caused by the movement. The distance of the moving object from the camera must also be considered. Distant objects require less shutter speed than near objects do. Objects moving directly toward or away from the camera require less shutter speed than those moving at right angles or nearly so.

Tone Separation

There are three outstanding faults with the photographs of beginners—lack of sharp focus, poor composition, and incorrect exposure. The technical quality of a photograph is judged by the clear-cut distinction between the light, dark, and middle tones. This distinction is called tone separation. Good tone separation means that the various light and dark parts of a picture correspond to those of the actual subject. Assuming proper development, tone separation depends almost entirely on exposure.

EXPOSURE METERS

Exposure meters, or light meters, are devices that accurately measure the light value of a scene or object, and provide a means of converting this information into usable shutter speeds and f-stops.

All exposure meters work on the same general principles; they differ only in appearance, as made by different manufacturers. The light energy is converted into electrical energy by a photoelectric cell. The cell is connected to a meter movement, which actuates a pointer needle. The needle indicates the light value of the scene or object to which the meter is pointed. A calculator dial assembly converts this reading into f-stops and shutter speeds. In order to obtain an accurate reading, the meter should not be pointed toward the light of the sky. It is also advisable to obtain an average between the lightest and darkest parts of the scene.

FILM

Types

Orthochromatic

Orthochromatic literally means "correct color." When film emulsions containing dyes were first manufactured, it was believed that maximum color sensitivity had been attained—that they were capable of rendering all colors in a corresponding shade of grey; thus, they were called orthochromatic. It was found, however, that although they were capable of recording greens, they were still not sensitive to red. In a photograph made with these films, blue and green are rendered as light tones, while red is reproduced as black. These emulsions react well to light produced by photoflash and are widely used for portraits and press photography.

Panchromatic

Panchromatic literally means "all colors." Some years after the perfection of the orthochromatic emulsions, dyes were discovered that would record all colors. Film made with these newer dyes is called panchromatic.

Panchromatic film covers all visible colors in varying degrees, depending on the dyes used in sensitizing the film. This film is manufactured in two types. The first has fairly even sensitivity to blue, green, and red, and is known as type B. The second is slightly more sensitive to red than to blue and green, and is known as type C.

When a red object is seen aganist a green background, the eye distinguishes between the two by their color. If a nonchromatic film is used to photograph a red object against a green background, no light will affect the film, and both the object and the background will appear as black. If orthochromatic film is used, the object will be black, the background light. If panchromatic film is used, both will appear in slightly different shades of grey. Because panchromatic emulsions give the most natural rendering of colors and brightness in shades of grey, they are used nearly exclusively in investigative photography.

Film Speed

Film is divided into three classifications according to the sensitivity of the emulsion to light, which is expressed in a numerical rating, known as the ASA Exposure Index, and is referred to as film speed. The three classifications are slow, medium, and fast. They have the following numerical groupings: Slow films have ratings up to ASA 100; medium films, from ASA 100 to 200; and fast films, from ASA 200 to about ASA 1,200.

Film speeds are related to each other in much the same manner as f-stops and shutter speeds. For example, if film X has a speed of ASA 50 and film Y a speed of ASA 100, Y is twice as fast, or sensitive, as X and requires half the amount of light to produce a satisfactory negative. Each time the film speed is doubled, it is equal to one f-stop in exposure. For instance, in the example given, if film X is exposed at f/8, then film Y would be exposed at f/11 to produce the same negative quality.

All manufacturers of film provide with the film a data sheet giving the film speed, flash guide numbers, and other information pertaining to that particular film. If this information is used and the instructions followed, good-quality negatives usually result. This information is particularly useful to an investigator who is infrequently required to use photographic techniques.

Contrast

Bright objects reflect a great amount of light and cause dense metallic silver areas in the image on the negative after development. Dark objects and shadows reflect little light and result in thin density in the negative. The difference in the density of the bright objects and the shadow areas is called the contrast of the negative, and it depends on three variables—strength of the developer, temperature of the developer, and length of time of development.

Development

When film is exposed, the action of the light causes a latent change in the silver halides of the emulsion. In order for the image to become visible, it is necessary to "develop" the film. The developer acts upon the silver halides, forming black metallic silver. Only those areas of the film that have been subjected to light are affected, and only to the degree to which they have been affected by the light.

After development, the film is placed into a fixing bath, which dissolves the unexposed silver halides and prevents any further change in the density of the image because of fading or straining. After fixing, the film is washed to remove the remaining chemicals, and then dried. Each chemical in the developer has a different function.

Developer

The type and ASA rating of the film being developed determine the developer to be used. The manufacturer's data sheet, found in each package of film, will give the recommended developer, processing time, and temperature for that particular film. These recommendations are observed because they produce the best results, and court testimony, if required, can reflect that the film was processed according to the manufacturer's recommendations.

Time and Temperature

The easiest and most universal method for development control is the time and temperature method. Prolonged development will cause excess density in the shadow areas and loss of detail in highlights. The rate of development is affected by the temperature of the developer. To aid in determining the proper development time, a time–temperature chart is supplied with the instructions packaged with the film.

FILTERS

Filters control the amount of light entering the lens of a camera. This entry of light is modulated by the color and density of the filter. Exposure must be regulated to compensate for the various light modulations. The manufacturer's data guide inclosed in each film pack or roll indicates the exposure filter factor for that particular film with various types of lighting.

Uses

The uses of the filters usually purchased with the speed graphic camera are specified in the following table:

Basic Filter Uses

Subject	Effect	Filter
Skin tones	Natural	Yellow
Wood	Emphasize grain	Red (dark wood), yellow (light wood)
Automobiles	Lighten to show dents, etc.	Use filter the color of car
Fabrics	Natural	Yellow
"	Show grain	Use filter closest to color of fabric
Indoor crime scenes (mostly wooden)	Natural	Yellow
Outdoor crime scenes (mostly foliage and ground)	Natural	Yellow
Outdoor crime scenes (mostly buildings and structures)	Contrast buildings, sky, surroundings	Orange, red (red buildings only)
Indoor crime scenes (wallpaper and linoleum)	Natural	Yellow
Fingerprints	Emphasize print against background	Use filter closest to color of background
Documents	Eliminate ink stamps	Use filter closest to color of ink
Sky haze	Eliminate	Haze filter

A filter used to enhance or add contrast does not eliminate any portion of the subject. When a filter is used that does eliminate a portion of the subject from the final reproduction, as in the last two items in the table, then the subject must be photographed with and without the filter. This is necessary if a print is to be derived that may be shown in court; the altered and unaltered subject must be shown.

INFRARED AND ULTRAVIOLET PHOTOGRAPHY

Police crime laboratories are usually equipped to provide specialized photography that is beyond the capability of the investigator in the field. These photographic services include infrared and ultraviolet photography, photomicrography, and photomacrography.

Infrared and ultraviolet photography will nearly always be done in the laboratory, and evidence requiring their use should be sent there for specialized treatment. However, the investigator should know what these techniques can do for him, and he should be capable of using them in instances calling for them.

Uses

Infrared photography is used for a variety of purposes:

1. Document examination
2. Differentiation or comparison of dyes, paints, or inks
3. Deciphering writing on charred documents, fabrics, and leather
4. Emphasizing scars, skin blemishes, and tattoos
5. Camera traps, using infrared film and flashlamps

Ultraviolet photography is used as much as infrared, and in some instances for the same type of examination. If infrared examination of a document reveals nothing, it is examined with ultraviolet. Common uses of ultraviolet are:

1. Document examination
2. Photographing fingerprints on multicolored surfaces
3. Photographing body fluids, such as semen, urine, etc.
4. Photographing "invisible" ink

When ultraviolet or infrared photography is used, a normal photograph is also made, for comparison purposes.

Photomicrography is used to examine minute detail and utilizes a camera mounted on a microscope.

Photomacrography is used to photograph small objects and uses a camera with the normal lens extended by a series of tubes, or a short-focal-length lens with a large bellows extension.

Techniques

Infrared

Infrared film is similar to other film in that it is sensitive to blue and violet light. It differs from ordinary film in that it is also sensitive to red light, which is visible, and to infrared light, which is invisible. Since infrared film is sensitive to blue as well as red and infrared light, it is necessary to use a filter that will absorb blue light. The Wratten filter No. 25A is standard for most infrared work; however, the Wratten 87, which excludes all visible light, is best for investigative infrared photography. Photographs can be made with infrared film without a filter, but they are not infrared photographs. They give about the same results as a photograph made with blue-sensitive film. All the light sources commonly used for normal photography are satisfactory for general infrared work.

Because of its longer wavelength, infrared light focuses at a plane different from that of visible light. The camera should be focused as though the subject were actually nearer to the camera than it is. The short focal length of the 35-mm cameras, with their greater depth of field, will normally allow proper focus without any compensation, provided that a small f-stop is used. However, the press- and view-type cameras will require some compensation. For a Speed Graphic with a 127-mm lens, the lens will have to be moved forward about 1 mm after focusing with visible light.

In general, dyes that appear light to the eye do not appear dark in infrared, but many dyes that are dark to the eye reproduce light in the infrared photograph. The difference in the infrared reflecting powers of two kinds of dyes in material may be important. The structure of the weave of dyed cloth often appears more clearly in photographs made by infrared.

Infrared is particularly useful in discovering writing or printing on a document that has been charred by fire. Documents that have deteriorated through age, or have acquired obscuring accumulations of dirt, often photograph successfully with infrared procedures. Infrared may also be used to reveal writings that have been erased or covered over with dark ink. (See Figures 4.1, 4.2, 4.3, and 4.4.) Using infrared as

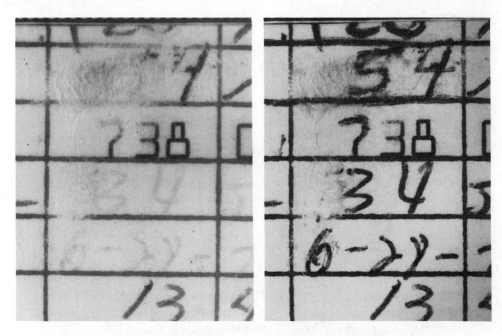

FIGURE 4.1 Panchromatic photograph, showing erasures

FIGURE 4.2 Infrared photograph with 87 filter, revealing erased figures

FIGURE 4.3 Panchromatic photograph, numbers covered with dark ink

FIGURE 4.4 Infrared photograph, revealing numbers

oblique lighting is particularly useful when mechanical erasure has caused the obliteration.

Inks, dyes, and pigments that are visually the same can often be differentiated by infrared photography. Fabrics that appear to be similar but have been dyed differently are frequently identified by using this technique. Infrared photography may also reveal the contents of sealed envelopes, and it is valuable for detecting stains on cloth, including bloodstains, that are not visible to the eye. Powder residues surrounding bullet holes in cloth, even when the fabric is dark in color or bloodstained, may be made visible by infrared photography.

Ultraviolet

Sunlight and tungsten lamps can be used in the reflected method of ultraviolet photography. The most common source of ultraviolet radiation, however, is a mercury-vapor lamp equipped with a filter, such as Corning Violet Ultra 5860, that transmits only ultraviolet light.

All films are sensitive to ultraviolet light and can be processed in the normal way. The use of ultraviolet requires lengthy exposures, so a film having a high speed should be chosen. The most sensitive of the human organs to ultraviolet is the eye; therefore, if exposure to ultraviolet light for prolonged periods is expected, protective goggles should be worn.

There are two methods that can be used with ultraviolet photography, one using the reflected ultraviolet light and the other using the induced fluorescence that is peculiar to some materials. In the reflected ultraviolet light method, no visible light is permitted to reach the film; a filter such as the Wratten 18A is used. In the second method, the light source must transmit only ultraviolet light, and the exposure should be done in a darkened room. It is necessary to use a filter that will absorb the photographically stronger ultraviolet light that is reflected from the object and transmit only the visible fluorescent light; the Wratten 2A is ideal for this purpose.

COLOR PHOTOGRAPHY

Type of Color Films

Color films come in two general categories: positive and negative. Positive color film produces a positive image known as a transparency (slide). Negative color films produce a negative image, which is then transposed into a positive print, positive transparency, or black-and-white print. Positive color films are distinctive in that, regardless of manufacturer, the brand-name ending is always "chrome" (as in Ektachrome, Anscochrome); for negative color films, regardless of manufacturer, the brand-name ending is always "color" (as in Ektacolor, Anscocolor).

Exposure

Color film is much more critical in its exposure latitude than black-and-white film is. Under- or overexposure will result in incorrect color rendition. Reflections and deterioration of dyes also produce false color values. The exposure of color film is calculated with the light meter in the same manner as for black-and-white films, except that the exposure is based on highlight readings rather than on shadow detail. Flash exposures are based on the guide numbers that are listed in the instructions for the film or flash lamps.

Use in Police Work

Negative color films for police work produce a negative that many copies can be made from. Also, if necessary, black-and-white prints and color transparencies can be produced from the same negatives.

Positive color transparencies, when viewed through transmitted light or projected on a screen, portray a faithful color rendition. They are another method of presenting evidence, since the entire court can see the same thing at the same time, and any explanation of details of the subject matter can be made easily.

Color prints produced for court use must be accurate color reproductions of the subject or crime scene. There must be no color shifts or extreme color impurities. Color shifts or impurities can occur when the laboratory technician printing the color was not at the crime scene and cannot accurately reproduce the colors without a color guide in one of the pictures. The investigator can aid the color-printing technician in one of the following ways:

1. Take one extra photograph of the crime scene or subject, and add a neutral density card, commonly known as a "grey card," to the picture. Neutral density cards can be obtained through photographic supply stores. The color-printing technician will have the same type of card in his possession when he is printing the pictures, giving him an accurate color to compare with the color on the neutral density card in the picture.

2. Enclose a small, colored piece of nonevidence from the crime scene with the negatives. The piece of nonevidence material should be shown conspicuously in the crime-scene photograph, so a color comparison can be made.

When presenting color photographs in court, always have black-and-white prints available in case the court does not accept the color prints.

PHOTOGRAPHING CRIME SCENES, AND OTHER USES OF PHOTOGRAPHY

Photographs taken at a crime scene can be studied later to find clues that may have been previously overlooked; and a photograph will always show more than a witness is able to recall with certainty. For these and other reasons, the most important rule is to photograph all evidence at a crime scene before anything is touched or moved. It is not expected that the scene of a crime will be left undisturbed for an indefinite period of time, so a record of the position and the condition of the details of the scene must be made before they are altered in any way.

In an investigative photograph—one prepared for the purpose of illustrating a point relative to a matter under investigation—care must be taken to make sure that no distracting object is near the center of the field of view or close to the camera lens, to detract from the main subject. Photographs should show clearly the point or points they are intended to prove.

A careful study of the image on the ground glass will enable the photographer to reproduce the scene as it appears to the eye. Objects and conditions that are not relevant to the crime scene should not be included, and unusual camera angles should be avoided where possible. However, photographs should be prepared to show the crime scene from all normal angles and include all evidence possible.

The photographs may explain how a crime took place and show all possible elements of the crime, so the investigator should proceed carefully, without undue haste, in order that errors will not be made that may affect the quality of the photographs.

Notes should be kept concerning the time of day the photograph was made, the type of camera used, the type of lens, the type of film, the diaphragm and shutter setting, the light source, the filter used, and the distance from camera to subject, as well as the height of the lens above the ground. All camera positions should be shown on the crime scene sketch.

When the crime scene is indoors, photographs of the house or building and the surrounding grounds should be taken. All rooms having a direct connection with the one in which the actual crime occurred should be photographed, as well as points of entry or exit, and places where evidence has been concealed. Photographs should be taken to show the situation, including evidence of a struggle, drinking glasses, food on tables, lights burning, and anything unusual or unnatural. A series of overlapping photographs should be taken in a clockwise direction around the room or from opposite ends of the room. Faithful reproduction of distance relationships is essential.

If the crime is outdoors, photographs should be taken that will identify the location. Impressions of footprints, tire tracks, and effects on foliage, such as bent grass, broken twigs, or branches, can be very valuable.

Investigators should not be photographed in the scene.

Homicides

In addition to those of the crime scene in general, photographs in homicide cases should be taken showing the location of the body and its position. At least two should be taken of the body, at 90-degree angles to each other, with the camera placed as high as possible, pointing downward toward the body. Closeup views of wounds can best be done at the morgue.

After the body has been moved, additional photographs should be taken of stains or other evidence located in the space that was occupied by the body.

Arson

Photographing a fire is usually divided into three phases—the fire in progress, immediately after the fire, and during cleanup operations.

Photographs of the fire while it is in progress should show the area of origin; speed, direction, and manner of spreading; progressive stages of the fire from various angles; and the arrangement of windows or doors. Photographs of spectators should also be taken, as possible suspects may be in the crowd. This phase of arson photography lends itself well to the use of color motion pictures. A time log should be maintained, recording the time each sequence was taken. If possible, photographs should be taken inside the burning structure, of any incendiary device found and any protective devices that may have been tampered with.

It is important to take as many photographs as possible, and as soon as possible, after the fire is out. Walls and chimneys may have to be pulled down as a safety measure, and this act might destroy valuable evidence. Special attention should be given to the arrangement of items in various rooms. Photographs should be taken of documents and records that appear to have been left purposely exposed, and doors, windows, and ventilators that appear to have been opened to provide cross-drafts. Photographs of electric clocks may provide information concerning the time of power failure. Deep charring and the small checks of the "alligator" pattern in burnt wood are best photographed with cross- or sidelighting. The exposure should be increased by one or two f-stops, as charred wood absorbs a great amount of light. Angles and lighting must also be taken into consideration.

During cleanup operations, photographs should include the complete exterior of the structure and surrounding area. Any evidence uncovered during the cleanup operation should be photographed in place.

Vehicle Accidents

The scene of an accident should be photographed as soon as possible after its occurrence. A large number of photographs should be taken to assure adequate coverage. The camera should be so placed that the entire negative will be utilized for objects or areas of evidentiary value. Prompt photographic action will indicate visibility and weather conditions, the exact positions of the vehicles, parts of the vehicles, and location of the victims. Photographs should be taken of points of impact, marks of impact, registration numbers, nature of the roadway, damage to real property, skid marks, tire marks, and glass or other debris on the highway.

Detailed photographs of the damage inflicted on vehicles are best taken after the vehicles have been removed from the scene. Closeups of special marks or identifications made by objects that struck the vehicle, or that the vehicle struck, are especially valuable.

Hangings

Overall views of the body should be taken from torso height, and should include the rope. Closeup photographs are taken of the victim's head and neck, showing the rope and knot, and another showing how and where the other end of the rope is secured. Any object that appears to have been kicked from under the victim's feet should be photographed, as well as any scuff marks that are near his feet.

If the body has been taken down but the rope is still secured, it should be photographed. If the rope has been removed from the victim's neck, closeup photographs should be made of the groove, including any black-and-blue marks or scratches. Appropriate photographs should be taken of objects that may indicate sexual deviations, such as chains, whips, female clothing being worn or not properly accounted for, and pornographic pictures or literature. In this connection, odd or bizarre ways of attaching the rope to the body should be photographed in detail.

Some evidence may be found on the body after it has been removed to the morgue. This may include teeth marks on the lips and tongue, protrusion of the tongue and eyes, bruises or scratches on the body, and marks on the arms and legs indicating that they had been bound.

Riots and Disorders

Investigators may be called upon to gather intelligence on such activities as riots and disorders. Motion pictures taken of participants aid in identification of those involved, and frequently lead to the arrest of leaders and agitators. Still pictures in black and white, as well as in color, are also of value for this purpose. In addition, such photographs serve as an excellent method for critiquing an operation and for planning future operations of a similar nature.

Surveillance

One of the primary purposes of a surveillance is to provide information and evidence concerning crimes and criminal acts. Surveillance photography is the same as other photography, except that the photographer must maintain secrecy. This means that the photographs must be taken with a telephoto lens or from a concealed camera position.

On a fixed surveillance, the prime considerations are concealment and field of view. A photograph should first be taken of the scene in general, to establish the location of the camera. If motion pictures are used, the first few feet of the film should be taken of a data board, and if a clock is located within the area or adjacent to it, the clock should be photographed at the beginning and end of each scene.

A moving surveillance is best adapted to photography with the use of the 35-mm camera. The investigator is out in the open and concealment presents some problem. Depending on the reasons for the surveillance, a telephoto lens may or may not be advantageous. If the moving surveillance is by automobile and photographs are to be taken, two men are required.

Camera Traps

A camera trap is a type of mechanical surveillance. Camera traps are used to photograph intruders or burglars. A camera is placed in a concealed position and loaded with infrared film. The room is darkened and the exposure made with infrared flash. A dim nightlight should be left burning to prevent the intruder from seeing the faint red glow of the flash lamp, should he happen to be looking at the camera when the flash is fired. The camera may be tripped by a mechanical device such as a trip cord, or by electrical means such as a pressure switch in the door or on the floor. It is essential that the field of view cover the trip area.

LEGAL PHOTOGRAPHY

Pictures for Detection and Evidence

The camera is a useful tool for investigating crime and for legal actions. Pictures are used as permanent records of the scene of a crime or a traffic accident. They are also used to compare and record bullet markings, fingerprints, signatures, and documents. By studying photographs, investigators may gain information that will lead to the solution of a crime, to apprehension of criminals, or to stolen goods. And photographs can be used as, or to support, evidence in court.

You can make fingerprints, certain dyes, and some invisible inks show up by using panchromatic film with a yellow filter and ultraviolet lamps. You can use infrared film to read faded documents and charred paper, to bring out old scars and tattoo marks, and to see through grease, grime, and some types of paint.

A Photograph Alone Is Not Proof

A photograph by itself is not admissible evidence in a court of law. Someone must attest to the authenticity of the picture. To aid in this, each picture must have a caption, and must be carefully protected. The caption must be detailed and exact. It should include names, places, dates, times, circumstances, and perhaps measurements, as well as photographic data such as focal length, camera angle, and film type. A sketch—even a rough sketch—of the scene will be helpful, especially if you make more than one exposure to cover the scene.

Legal Photographs Are Factual

A legal photograph must be clear, sharp, and undistorted. The intent is to picture the facts as they are, not to capture impressions.

Do not use dramatic lighting and other photographic tricks that improve the eye appeal and stress the concepts of regular photographs.

Show scenes in their natural state. Don't add or remove anything. Carefully note any exceptions that are due to actions of police or others. To provide a natural view, photograph scenes from eye level with a normal-focal-length lens. Use a tripod or other solid support to steady the camera so that the pictures are sharp.

A fine-grain film is recommended for legal photography, because of its capability for small detail and clear enlargement. Enlargement may be necessary for both the investigative and presentation stages of legal action, to clearly show some small visual fact. Even without enlargement, small detail is desirable for sharpness and clarity.

Use panchromatic film unless color is an important factor, and then use color film. Use infrared film when its ability to detect difference in material or to see through substances is required.

5

Casting and Molding

INTRODUCTION

This chapter provides information on the protection and preservation of various impressions discovered at the scene of a crime, and on some of the materials that may be used and procedures to follow in making casts and molds. It includes the methods for making casts and molds for identification purposes, various suitable ways for packaging such evidence materials for shipment to police crime laboratories, and the basic information used at the police crime laboratory in the evaluation of casts and molds.

The examination of a crime scene will frequently reveal footprints, tire prints, toolmarks, and other impressions that, if properly preserved and evaluated, may provide valuable investigative leads. Casting and molding is the process used to reproduce impressions found in such locations, or on such bulky objects, that their removal to the laboratory or courtroom would be impossible or impractical. The evaluation of casts by comparison with objects thought to have made the impressions often provides a positive means of proving that certain persons, tools, or tires were at the scene of the crime. Evidence of this type is especially fragile, subject to the action of the elements, and to obliteration or partial destruction by human carelessness or design.

Products manufactured by the same factory may have the same general appearance but differ radically upon close examination. Even crepe rubber soles, whether made from unvulcanized rubber or by calendering, will seldom have the same design. The cutting of a sole will not be duplicated exactly by any subsequent cut. Molded materials such as tires and rubber heels will require the finding of individual characteristics before a positive identification can be made. When insufficient detail is present to enable a positive finding, general similarities in size, design, and shape may be important circumstantial evidence.

DEFINITIONS

The following terms are used in this chapter:

Mold: A negative impression. Details appear reversed when viewing the object and the mold together.

Object: Anything making a mold, cast, impression, or mark of any description.

Cast: A positive impression made from a mold.

MATERIALS REQUIRED

The following materials, commonly used in casting and molding, are available through commercial police equipment supply houses:

1. Plaster of paris
2. Rubber bowls and enamelware (Never use metal bowls.)
3. Spatula
4. Lacquer, colorless
5. General-purpose thin-weight oil
6. Shellac
7. Syringe
8. Talcum powder (for casting in snow)
9. Commercial casting and molding compounds
10. Ruler

11. Reinforcing materials, such as fine-mesh chicken wire, wire screen, grass, gauze, coat hangers cut to desired size, twigs, or other small pieces of wood

12. Framing materials such as earth, for building around footprints or tire prints; or linoleum, lead strips, metal venetian-blind slats, or wood

PRESERVATION AND RECORDING OF IMPRESSIONS

Preserving

When an impression is located at a crime scene, the first thing to do is preserve it from destruction. Footprints or tire prints found in any medium, indoors or out, may be covered with anything available, such as cardboard boxes, ashcans, garbage cans, or wastebaskets. If outdoors, the general area may be staked out and marked with tape. The area should be large enough and the boundaries established at a sufficient distance to preclude the obliteration of impressions that may still remain to be discovered.

Photographing, Sketching, and Recording

Photographs of the impression should be taken first from a distance, to show its relationship to other objects at the crime scene. Closeup photographs are then taken. The first of these should be made with as little preliminary preparation of the impression as possible. The camera must be placed so that the plane of the film is parallel to the ground. The shutter should be set at the smallest permissible f-stop in order to obtain the greatest depth of field. A flash should be used, even in sunlight, to make sure that the details are clearly defined. The flash should be held close to the ground, as the resulting oblique light will reveal more details. Another closeup photograph should then be taken with a ruler placed alongside the impression, so that the proper scale of the photograph can be determined. If a paper evidence rule is not available, identifying data such as date, case number, and investigator's name may be written on a piece of paper and included in the photograph for future reference. (See Figures 5.1 and 5.2.)

The location of the impression is recorded in the sketch of the crime scene, with appropriate measurements. When a portion of a tire print is to be cast, the location is most important. Distinctive characteristics such as trademarks, nail holes, cuts, scars, and other imperfections should be

FIGURE 5.1   Overall measurement of shoe print, with identifying data

FIGURE 5.2   Sole measurement

drawn and included in the sketch. Measurements and other descriptive data should be included in the investigator's note. Footprints should be measured, without touching the impressions, to determine the overall length, length of the heel, width of the heel at its front edge, and width of the sole at the ball of the foot. If outstanding characteristics are observable in a tire print, the distance between two of these adjacent characteristics will indicate the circumference of the tire.

CASTING

Preparing the Impression

When a print is found indoors in dust, or if it is made only from dirt adhering to the object, no preparation may be required. An impression found outdoors should be examined to determine if any loose particles have been blown or otherwise introduced into it. Any such particles should be removed with extreme care. A syringe may be used to gently blow away dust that may have gathered, a pair of tweezers to lift out small stones, or a pocketknife to flick out such debris, but care must be exercised to avoid destroying portions of the print. The syringe can also be used to good advantage to withdraw any water or other liquid that may have accumulated in the impression. Water can also be drawn off by cutting a small channel at one side of the impression to allow drainage.

Plaster of Paris Casts

Plaster of paris is especially suitable for taking impressions of foot and tire prints or other marks in dirt, mud, sand, or snow, if they do not require microscopic-detail accuracy. Where such detail is required, a casting medium other than plaster must be selected.

Preparation before Casting

In sandy soils, the particles frequently lack cohesion, and the print is therefore fragile. The impression should be strengthened to support the weight of the casting material and to prevent the destruction of fine detail. This is best done by spraying the impression with a plastic spray or shellac; but hair spray makes a good substitute if these are not immediately available.

The application of a direct spray may cause destruction of the de-

FIGURE 5.3
Spray directed at baffle

FIGURE 5.4
Retaining wall made from available material

tails, so the spray should be directed against a baffle made of cardboard or any other suitable material, to cause a fine mist to settle gently into the print. (See Figure 5.3.) The number of coats to be given can best be determined by an examination of the test print. Usually, from three to ten coats are sufficient in sandy soils. Shellacs, varnishes, and plastics are used only to solidify the print and are not always essential to the casting process.

When the spray has completely dried, a thin mist of light oil should be allowed to settle in the print, making it easier to remove the completed cast. If the print is in a good solid earthy material, no further preparation is necessary. A retaining wall should be placed around the print to confine the plaster to a convenient area and to allow the cast to be built up to the desired thickness. The retaining wall may be constructed of any available material, including earth, to build a damlike structure. (See Figure 5.4.) If strips of material are assembled for this purpose, and it is intended that they be used over again, they should be given a light coat of oil prior to use. This will allow them to be more readily removed from the cast.

Mixing the Plaster

The plaster of paris mixture may be prepared by one of two methods, both of which require the sifting of the plaster into water. *Never add the water to plaster* (Figure 5.5). One method is to sift the dry plaster slowly into the water while constantly stirring the mixture. In the second method, the plaster is added to the water by sifting it around the edges of the container until it begins to rise to the surface of the water. When cracks like those seen in dried mud appear, no more plaster should be added. Mix by stirring beneath the surface to a thick creamy consistency. If lumps appear, they should be removed.

It is advantageous to make a thin mixture to place in the impression first, to record the finest of detail, since it can flow freely into the impression, and then follow this with a thicker mixture. Care must be taken not to make the first mixture too thin, as it might wash away details. Experimentation will best indicate just the right proportions. If a mixture is made too thick, it must be discarded and a new batch mixed. The mixture is ready to pour when it has reached the consistency of thick cream.

Pouring and Strengthening the Cast

The plaster should be introduced into the impression at a place where very little if any detail exists. It should be poured from a low level, and the force of its falling into the impression should be broken by letting it drop first onto a spatula or the hand, moved in a circular motion to cover the entire area of depression. (See Figure 5.6.) These precautions are taken to prevent destruction of important details.

After a layer of plaster about ¾ inch thick has been poured, the reinforcing material should be added. If sticks or wire are used, care must be taken to prevent the ends from protruding through the bottom of the cast. If dried twigs or wood are used, they should first be soaked in water, as they will absorb water from the cast, making it more fragile. Twigs, sticks, and pieces of wire should be placed at random in the cast, not laid parallel. If the reinforcing material is laid parallel in a single direction, the cast may fracture in the plane in which the pieces have been laid. Wire mesh does not present this problem.

After the reinforcing material has been introduced into the cast, the rest of the plaster can be added until the desired thickness of the finished cast has been reached.

FIGURE 5.5 In preparing casting mixture, do not add water to plaster

FIGURE 5.6 Mixture in retaining wall, approximately ¾″ thick

Additives

Although a thin mixture of plaster of paris will record more fine detail than a thick mixture, it will require a much longer setting time. One-half teaspoon of salt added to each pint of water used in mixing the plaster will hasten the setting. The more salt, the faster the setting. Sugar added to the water will retard the setting. A saturated solution of borax—one part borax to ten parts water—will retard the setting from 15 to 30 minutes, and also make the impression harder and sharper. Substances used to hasten or retard the setting should be added to the water before the plaster is added.

Special Casting Situations

Casting under Water

It may be necessary to cast a footprint or tire print that is entirely under water, the location and amount of which preclude its draining and removal. A section of stovepipe may be used to direct dry plaster to the print location and prevent wasting the plaster. The dry plaster should be sifted into the stovepipe and directed into the impression. This method can be used even in fairly deep water. Three to four parts of salt may be added to ten to twelve parts of plaster of paris to hasten the setting time.

Casting in Snow

The temperature, depth, and adhesive quality of the snow and the condition of the ground surface are all-important factors. Plaster of paris emits some heat as it hardens, and this might melt the snow, destroy the print, and impair the details of the cast. The investigator should make trial casts in the snow away from the impression, to determine which of the several techniques is best in the situation. The imprint may be strengthened by spraying with plastic spray, then have a thin layer of talcum powder dusted into it to act as insulation against the heat. The impression should again be sprayed with the plastic. Several coats of spray may be necessary to suitably fix the impression.

A retaining wall is especially necessary here, to prevent the spread of the plaster to areas not insulated, in which case the heat of the setting plaster might melt the snow and permit water to run under the prepared imprint, further contributing to its possible destruction.

Completing the Cast

Identification

Before the cast is completely set, it must be properly marked for identification. The identifying data can be scratched into the surface of the cast with any suitable instrument. As a minimum, the data should include the case number, date, investigator's initials, and any other information deemed necessary. An arrow indicating the direction of north will enable the investigator to locate the exact position of the cast in relation to other evidence found at the crime scene. When several casts are made at the same location, they should be numbered consecutively, with the number of each cast entered in the investigator's notebook, along with data describing the original location.

To assist the laboratory technicians with their examinations, a number of casts may be taken of different shoe and tire impressions found at the crime scene. Often, a print made by a particular shoe will clearly indicate details and characteristics not found in a second print made by the same shoe.

Removing the Cast

After the cast has completely hardened, it may be removed for further processing. Plaster usually hardens within 20 to 30 minutes after preparation. During the process of setting, the plaster becomes warm; when it starts to cool, it has hardened sufficiently for removal.

Care must be exercised in its removal; even though it has been reinforced, plaster is still fragile. After removal, any excess dirt may be dislodged by washing the cast gently in a pan of water, with a hose, or under a faucet. If a hose is used, the nozzle must be removed, since water under pressure may round off edges or obliterate small details. Use of a brush or water under pressure should be avoided in order to prevent possible damage to details in the face of the cast.

If portions of the surrounding soil are to be collected for petrographic comparison with the soil adhering to the shoes or clothing of a suspect, it may be appropriate to send the cast to the laboratory without washing. This will provide the laboratory with samples of the soil directly under the shoes of the suspect at the time he made his imprint, and may be of aid to the technicians in making a comparison of the soils. The same is true for tire imprints. (See Figure 5.7.)

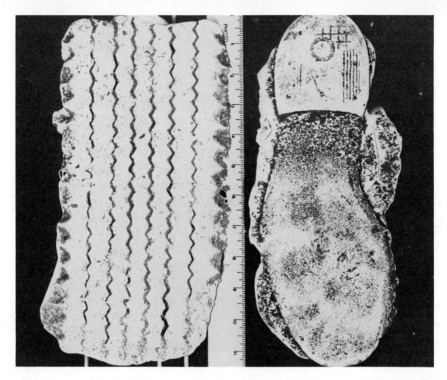

FIGURE 5.7   Dirt samples contained in shoe and tire imprints

Other Casting Materials and Possible Uses

Silicone Rubber

Silicone rubber, although more expensive than plaster of paris, offers some distinct advantages: Casts can be made from it quickly; water or heat is not needed; and castings are rubbery, eliminating the problem of breakage during handling or shipping. The rubber freezes at extremely low temperatures, making it very useful in casting tracks or prints in snow. It is a fast-cure, room-temperature vulcanizing rubber, requires no heat, and sets up into a solid rubbery mass. The catalyst (a substance that causes or accelerates a chemical reaction) is supplied in a small tube with the silicone-rubber package. The catalyst should be thoroughly mixed with the liquid rubber just before using. About one-half teaspoon of catalyst to a pound of the rubber is normal. The curing time of the rubber can be varied by altering the amount of catalyst added; a table is supplied with the material. A curing time of five to ten minutes for prints in dust is recommended.

Care must be taken when mixing to preclude the formation of air bubbles, as they may obscure details in the cast. When catalyzed, the rubber will remain workable for about five minutes at 77° F., after which it will set. Lower temperatures lengthen this period.

Suggested uses include foot and tire prints, dust prints, tool and jimmy marks, casting of parts of the human body, and fingerprints. Silicone rubber is not recommended for surfaces bearing natural patterns, such as leather or fabrics, as the detail of the print is obscured by the detail of the natural surface. Epoxy casting resin may be used to make a positive from the silicone-rubber impression.

Latex Rubber

Latex rubber has been used successfully to reproduce very faint impressions on linoleum, and for fingerprints developed with powder. However, the impressions on the latex rubber have proved to be fleeting and without permanent stability.

Liquid Sulphur

Sulphur gives very fine details, but it is not as readily available as plaster of paris. The retaining wall should be greased with a light coat of oil. If sulphur is to be used to cast a toolmark, the object on which the toolmark is present should also be lightly greased. Clark recommends a formula of eight parts melted sulphur and one part iron filings. This mixture is allowed to cool and, while still liquid, poured over the object to be cast. Gently blowing the liquified sulphur will propel it into the smallest of indentations.

Sulphur transported to the scene in a thermos bottle may be used for casting in snow. The liquid sulphur crystallizes on contact and gives excellent detail.

Moulage-Agar Compositions

High-quality reproductions, similar to those described under silicone rubber, may be made of complicated impressions by using colloidal substance of the compound agar to make the negative cast. The moulage, or negative compound, in chunk form, is heated in an enameled double boiler until the material is smooth and the consistency of thick batter. If it is too thick, water may be added to thin it. The heated compound may be carried to the scene in a thermos bottle or other insulated container.

Retaining walls and other surfaces require no special preparation, as these compounds will not adhere to them. The first coat of moulage should be applied directly on the object with a brush or spatula. Care must be taken to brush the moulage into all corners and irregularities. Air bubbles are prevented by vigorously brushing on the first coat. Subsequent layers should be applied while preceding layers are still soft.

Coarse cheesecloth or wire screen may be used as reinforcement and additional moulage brushed on over the cloth or screen.

These compositions are good heat conductors, and while the surface may appear to be set, the cast may still be soft inside. Ice may be applied to speed setting. After it has fully cooled, the cast may be carefully removed. Where there is an undercut, it may be necessary to slit the mold and subsequently close it after removal.

Special Techniques

Footprints

When a search indoors is to be made for the purpose of locating footprints, it is advisable to darken the room first. A flashlight, with the beam aimed low, should be used to search floors, window sills, and furniture. Oblique lighting will often show up traces that are not visible with ordinary or direct light. Footprints on carpets should not be overlooked, as appropriate photographic techniques may produce valuable evidence photographs. Excellent results have been obtained by using a high-contrast film, and a high-contrast paper for the print.

Footprints on a solid surface, made by the dust particles adhering to a shoe, can be lifted by the use of large sheets of ordinary fingerprint-lifting tape. Another method is to use photographic film prepared in the following manner: Unexposed film is fixed, washed, and dried (this may be prepared in advance and kept until needed for use); the clear photographic film is moistened by soaking for a few minutes in water; the excess water is squeezed from the film, and the film applied to the dust print with a roller. Soaking the film in water makes the emulsion side tacky, and the dust particles will adhere to it. By using clear photofilm, the laboratory technician can choose backgrounds of varying colors to aid him in obtaining the best possible contrast for comparison purposes.

Toolmarks

Toolmarks on metal surfaces, such as hammer blows on a safe, are usually extremely faint and the detail microscopic. The wax portion of positive moulage material (see below, "Positive Moulage-Agar Compositions") is excellent for use in these cases. Marks having deeper indentations may be cast using any of the materials previously mentioned. These types of impressions, including jimmy marks on doors and window sills, are not subject to easy destruction, and other substances may also be employed: Modeling clay may be pressed into a toolmark for an excellent cast producing the fine detail necessary for positive identification; tissue paper macerated in water and mixed with glue will reproduce a toolmark in the form of papier mâché.

Toolmarks may be destroyed or changed if tampered with by an untrained or unskilled investigator. Original evidence is more useful for scientific examination and evaluation and is less subject to attack in court than reproductions are. The investigator is often unable to make photographs and casts that represent the evidence sufficiently for identification purposes at the laboratory.

Tire Impressions

The circumference of a tire is between five and eight feet, and the probability of identification or matching of a tire track with a particular tire increases with the length of the cast or casts obtained. Ideally, consecutive tire casts should be made equaling the circumference of the tire involved. If more than one track is found, casts should be made of each individual tire track. If it can be stated that the combination of the designs taken from a set of four tire impressions found at the crime scene corresponds to the designs and the wheel positions of the four tires on the suspect's automobile, such testimony is of obvious value.

Each cast and tire submitted should be fully identified as to the wheel position in sketches and photographs.

MOLDING

It may be desirable to produce a mold from a cast that had been made previously—from a footprint, for example. It might be advantageous to provide several investigators with copies of the cast so that they may search independently for the shoe that made the original impression. Once a mold is made of the cast, as many copies as are needed can be reproduced. Casts made with agar-moulage compositions require that a mold be made shortly after the cast, in order to meet evidentiary requirements.

Plaster of Paris

To make a plaster of paris mold from a plaster of paris cast, the cast should first be given a thin coating of light oil; otherwise, it will be most difficult, if not impossible, to separate the cast from the mold. With this precaution having been taken, the method for producing the mold is almost identical to that used when making the cast. The plaster of paris is mixed and poured into a container or onto a flat surface within the area contained by framing material. A rubber photochemical tray of appropriate size makes a good container. The cast, after being coated with oil, is placed into the plaster mixture, to an appropriate depth. When the plaster mixture has set, the cast is removed, leaving a mold that may be used for making other casts. If the duplicate casts are to be made of plaster of paris, the mold must be coated with oil, or the same problem of separating the cast from the mold will arise once again.

Positive Moulage-Agar Compositions

Positive moulage material, a wax, may be melted over a slow fire in any pan from which it can be poured. If the cast is hollow, the melted moulage should be poured in, allowed to stand two or three minutes, and then poured out. This procedure is then repeated as necessary, allowing a slow buildup of wax to the required thickness. The moulage may be brushed or poured into other than hollow negative casts. Since it will not adhere to anything, it may be used for making molds from plaster casts with no oiling of the cast. The cast can be appropriately framed and the material brushed onto it, or it may be applied with a spatula. The mold should be reinforced with wire mesh or cloth; bandages make excellent reinforcing material.

FORWARDING TO LABORATORY

Packaging

Even with reinforcing, a plaster of paris cast is fragile evidence and must be handled carefully. It should be wrapped in soft paper or cotton to avoid abrasion, friction, and possible destruction of fine identification points, then well wrapped in strong wrapping paper and fastened securely. It may then be placed in a box, cushioned on all sides by excelsior or other shock-absorbing material. Wax and modeling-clay casts, while not as fragile as plaster, present other problems. They are more easily deformed by pressure, so they must be protected from pressure as well as abrasive action.

Photographs of the impressions from which the casts were taken should be forwarded with the casts to the laboratory to aid the technician in his analysis. Articles such as shoes or tools that are to be shipped to the laboratory for comparison purposes should be boxed and shipped separately, protected from the accidental addition of marks that might void them as evidence. (See Figure 5.8.)

FIGURE 5.8
Cast and suspect footwear submitted for comparison

LABORATORY EXAMINATION AND ANALYSIS

Examinations of this kind of evidence are usually done in the fingerprint division of the laboratory, because of the close similarity of the types of examination.

The laboratory technician will examine the evidence to find identical individual characteristics. *Class characteristics* are those marks, protuberances, and lettering that are the direct result of the manufacturing process. *Individual characteristics*, which are the basis for a positive identification, are those markings accruing to the object in its daily use. They may be cuts, bruises, tears, signs of irregular wear, or other accidental markings that set the object apart from those of any other of like manufacture. The appearance of several of these kinds of characteristics in identical places, and in the same relationship to one another, would be so conclusive as to preclude every other finding except that the impression was made by the object with which it was compared.

Even if the laboratory technician is unable to find a sufficient number of identifying points to make identification, the class characteristics may still assist the investigator, in that those persons whose shoes, tools, or other objects do not match the class characteristics of the crime scene impression may be eliminated as suspects.

6

Fingerprinting

INTRODUCTION

This chapter is designed to assist the investigator in applying correct fingerprinting techniques and processes in the field. Assistance required beyond the information contained here requires more technical considerations; the investigator should seek such help from the criminal investigation laboratory, from people who are qualified as fingerprint classifiers with the Federal Bureau of Investigation, or from local civilian police organizations that have personnel who are expert in this area.

CHARACTERISTICS OF FINGERPRINTS

Fingerprints are the most positive means of identifying people. The ridges on the skin of the palmar surfaces of the hands and the plantar surfaces of the feet, commonly referred to as *papillary* or *friction ridges*, form on the fetus before birth and remain unchanged throughout life, and even after death, until decomposition of the skin destroys them.

Damage to the skin during the life of a person may be either temporary or permanent. Abrasions and slight cuts that do not permanently affect the skin are corrected in time by nature, and the ridges reappear as they existed before the damage occurred. Deep cuts and injuries affecting the innermost sections of the skin will result in permanent scars, but the general pattern will continue to exist.

DEFINITIONS

Bifurcation: The forking or dividing of one ridge line into two or more branches.

Core: The approximate center of the finger impression.

Delta: The point on the first bifurcation, abrupt ending ridge, meeting of two ridges, dot, fragmentary ridge, or any point upon a ridge at or nearest to the center of divergence of two type lines, located at or directly in front of their point of divergence.

Divergence: The spreading apart of two ridges that have been running parallel or nearly so.

Node: Protuberance found at the base of the fingers on the palm of the hand.

Palmar: Pertaining or corresponding to the palm of the hand.

Pattern area: That part of a loop or whorl in which appear the cores, deltas, and ridges used to classify a fingerprint.

Plantar: Of or pertaining to the sole of the foot.

Ridge: Elevated portion of the skin found on the palmar and plantar surfaces. The sweat pores run in single rows along the ridges and communicate through the sweat ducts with the sweat glands, which are below the entire epidermis. The friction ridges result from the fusion in rows of separate epidermic elements.

Shoulders (of a loop): The points at which the recurving ridge definitely turns inward or curves.

Type lines: The two innermost ridges, which start parallel, diverge, and surround or tend to surround the pattern area.

FINGERPRINT RECORDS AND FORMS

The basic forms for fingerprinting are:

1. Police Record Check

122 Fingerprinting

2. FBI U.S. Dept. of Justice Fingerprint Form, FD Form (Applicant)

3. FBI U.S. Dept. of Justice Fingerprint Form, FD Form (Arrest)

Fingerprints taken for record checks are placed on FD Form 249. In the "Charge or Offense" block, the word "Inquiry" is used. The cards are forwarded without letter of transmittal to Director, Federal Bureau of Investigation, U.S. Dept. of Justice, Attention: Identification Division, Washington, D.C. 20032.

EQUIPMENT

Equipment for the taking of fingerprints is available through normal police supply channels and also from commercial police supply houses. (See Figure 6.1.) Basically, the equipment consists of a device to hold the fingerprint card, a means to apply a suitable coat of ink to the fingers, and an appropriate ink. A piece of plate glass is ideal for smoothing the ink in a thin film for inking the fingers. Small dabs of ink are rolled out with a rubber roller. It is advantageous to place a strip of plain white

FIGURE 6.1   Fingerprint equipment

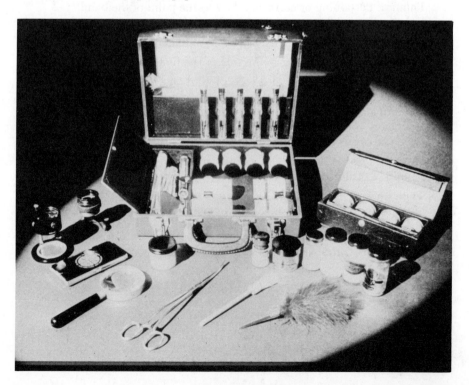

paper under the glass; this permits both a rapid evaluation of the thickness of the ink film and a check of the adequacy of the inking of the fingers.

Printer's ink or special fingerprint ink is recommended. This ink dries on the card quickly, reducing smudging, yet will maintain its consistency to preclude constant cleaning of the glass plate. There is also a commercial product available that resembles an office stamp pad, remaining "permanently" inked and ready for use and obviating the necessity of using glass, roller, and ink.

RECORDING FINGERPRINTS

The art of taking good, clean impressions is not difficult; it is a matter of knowing what constitutes a fingerprint and what the essentials of a pattern are, exercising a reasonable amount of diligence, using appropriate equipment, and keeping it clean. Fingerprint classifiers often remark that if the person who took a set of fingerprints was responsible for their correct classification, print quality would improve spectacularly.

The color of a good fingerprint impression should be dark grey. The ink must not fill in the depressions between the ridges. Practice is all that is required to produce professional-looking fingerprint cards.

The first step in obtaining a set of fingerprints is to have the subject sign the card, to establish his identity positively. It is recommended that this be the extent of the initial processing of the card. Next, have the subject wash his hands, in your presence, to remove perspiration and dirt. In drying them, care must be taken to ensure that lint from the towel is not adhering to the fingers. The subject is then fingerprinted, after which the person taking the prints completes the other required information on the card.

The front of the fingerprint card is divided into two main sections. The upper section is for the "rolled" impressions and the lower is for the "plain" impressions. The rolled impressions provide the entire surface of the fingers and thumbs, from nail edge to nail edge, and from just below the tip of the finger to about ¼ inch below the first joint. This large area provides all the necessary fingerprint ridge characteristics for proper and correct classification. Since the ten rolled impressions show more of the ridge structure and pattern area than the plain impressions do, they are more important.

Rolled Impressions

If you hold your arms in front of you with the backs of the hands touching, you will see that this is a strained and somewhat awkward posi-

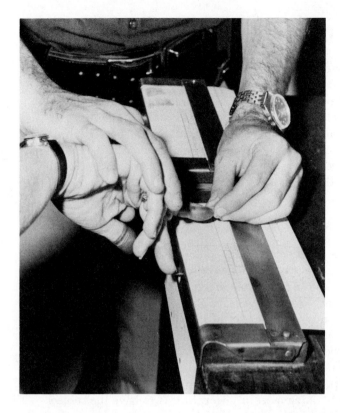

FIGURE 6.2
Starting the roll of the right index finger

tion. As you turn the hands over so that they are palm to palm, you will find that they are now in a comfortable position. The fingers are rolled by starting in the awkward position and ending with them in the comfortable position, or rolled *away from the center* of the subject's body. (See Figure 6.2.) The person taking the impressions should not exert a lot of pressure on the finger, but just enough to provide a clear, even, unsmudged print. A subject who is trying to help is usually more of a hindrance. He should be told to look away from the operation while his prints are being taken; he will be more relaxed, and a better print will result. The person taking the fingerprints can remove the finger from the card with the least chance of smudging if the subject is relaxed and the finger is in the position of least strain at the completion of the rolling process.

The foregoing method does not apply to the thumbs. If you hold both hands in front of you, palms up and thumbs extended, you find that there is a feeling of some strain. Notice also that the inside edges of the thumbs are down in the proper position for starting the rolling process. Now hold the hands in front of you with the palms down; you will see that it is a more comfortable position and that the outsides of the thumbs are in perfect position for completing the roll. Therefore, the thumbs are rolled *toward the center* of the subject's body.

Plain Impressions

The plain impressions should be recorded on the card at a slight angle, to ensure that ¼ inch below the first joint of the little finger is shown. The plain impressions are used to verify that the rolled impressions are in their proper order and to enable the classification technician to check on the appearance of certain characteristics that are sometimes slightly distorted, owing to the flexibility of the flesh of the fingers during the rolling process.

The plain impressions are made by having the subject hold his fingers together straight and stiff. The hand and wrist form a level plane with no break at the wrist. The person taking the fingerprints holds the subject's wrist and impresses the fingers simultaneously in the appropriate space. (See Figure 6.3.) The thumbs are also held straight and pressed in their respective places separately.

Inking

The procedure for inking the fingers is the same as that described for taking the prints. Never roll a finger back and forth on the inking slab to get enough ink; get enough the first time. If by accident you

FIGURE 6.3
Plain impression

Photographic Print Impressions

For taking prints where the fine structure is needed in detail for comparison, a good method is to use glossy photographic print paper and make the print with photographic developer. The developer is applied lightly to the fingertip and rolled lightly on the paper. The paper is placed in an acid fixing bath for approximately 30 minutes and washed in the same fashion as an ordinary photograph. This process yields a highly detailed print.

Palm Prints

In many cases, palm prints must be obtained to compare with palm prints, or portions of them, found at a crime scene. A discrepancy that is most frequently encountered in the taking of palm prints is the failure to exert sufficient pressure on the back of the hand to ensure the printing of the hollow of the palm and the nodes at the base of the fingers. If pressure is not successful in obtaining prints of these portions of the palm, a satisfactory print can sometimes be obtained by wrapping paper around a bottle and having the subject grasp the bottle in the manner best suit-

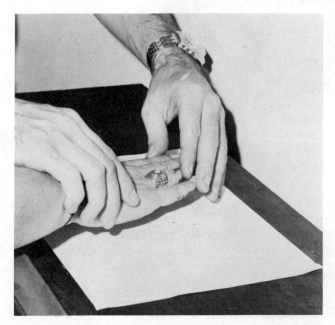

FIGURE 6.4
Taking palm print

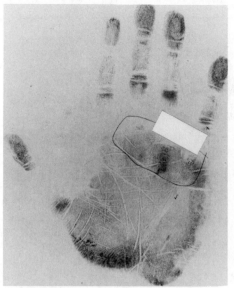 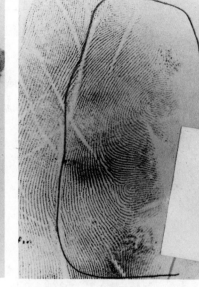

FIGURE 6.5 Overall handprint   FIGURE 6.6 Closeup of palm print

able to produce a complete print; or the palm can be rolled onto the paper by starting with the fingertips and having the subject roll the bottle along a table top, although some distortion might be introduced into a print so obtained. Also, when taking palm prints, it will be noted that only the side of the thumb appears; the whole thumb should be printed separately. (See Figures 6.4, 6.5, and 6.6.)

PROBLEM PRINTS

Deformed Fingers or Hands

If the fingers or hands are so deformed that the normal printing operations cannot be performed, other procedures will have to be attempted. Success has been achieved in inking the fingers by rolling out a thin film of ink on a spatula and transferring the ink to the fingers. It might be necessary to cut a square from another card and obtain a print by holding the square in the palm of the hand and rotating the square around the finger. A satisfactory print has been obtained by holding the square imbedded in modeling clay. When the best print possible has been obtained, the square can be taped or stapled in the proper finger space on the card and a notation made on the back of the card of the reason for this procedure.

In cases where amputations have occurred, a notation in the space or spaces for those fingers must be made. Examples: "AMP 1st joint Feb. 1973"; "Tip AMP."

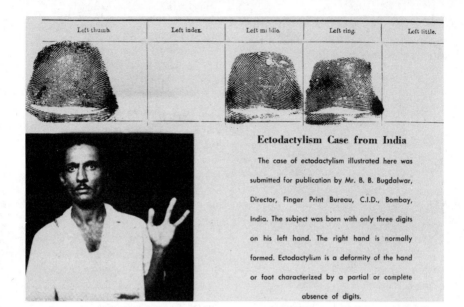

FIGURE 6.7   Case of deformed hand

In the case of extra digits, special care and notations must be made. Usually, where extra fingers occur, it is the little fingers or thumbs that are involved. In these cases, the inner five fingers are printed on the front of the card and the extra fingers on the outside of the hand are printed on the back, and appropriate notation is made.

In the case of webbed fingers, as much clarity as can be obtained, with appropriate notations that they are joined, must suffice. Any deviation from the normal fingerprinting procedure should be explained on the back of the fingerprint card. (See Figure 6.7.)

Dry or Soft Skin

As a result of certain kinds of occupational work, the skin on the tips of the fingers may become rough and dry. Rubbing the bulbs of the fingers with Vaseline or oil will often make them sufficiently pliable and soft to be printed. Ice held against the fingers will sometimes help, if the ridges are fine and small and the skin is soft, as with women and children.

Deceased Persons

Fingerprinting deceased persons is considered under three categories. First, a person recently dead, and before rigor mortis has set in, presents little in the way of real problems. For ease, place the body face down with the arms extended over the head. It is usually easier to roll the inking plate around the fingers than it is to roll the fingers. It is easier too if the card is cut either into strips, one for the right and one for the left hand, or into squares, one for each finger and thumb. Care must be exercised that there is no mixup of the prints when placing the squares on the card as the print is rolled.

The second type of cadaver, and one usually a little more difficult to print, is one in which rigor mortis has set in. In most cases, the fingers will be partially clenched, curved toward the palm of the hand. Pressing on the middle joint of the finger toward the palm can usually straighten the finger enough to permit inking and printing. Methods used and procedures adopted will be determined by the availability of equipment and the ingenuity of the person doing the fingerprinting.

The third, and by far the most difficult, is the case where the flesh has decomposed to some degree. It may be possible to have the autopsy surgeon cut the skin from the finger. The skin can be placed over the finger of the technician wearing rubber gloves, and inked and printed as though it were his own. The fingers may be dusted with bismuth or lead carbonate and X-rayed. If it is considered absolutely necessary, the fingers may be amputated. Each finger should be placed in a bottle containing distilled water or alcohol, making sure that each is properly identified. If the skin is shriveled, it may be expanded by injecting air, glycerine, liquid paraffin, or gelatin into the bulb of the finger with a hypodermic syringe. If the skin is dehydrated, the finger may be soaked in a 3 percent solution of potassium hydroxide in warm water for a short period of time, with intermittent washing, until the finger returns to normal size. If the skin is burned, photographing with infrared light may be successful, or it may be necessary to seek the advice of a laboratory technician.

LATENT FINGERPRINTS

The word *latent* means "hidden," or "not visible or apparent," but it has no such limited application to the fingerprint technician. It should be understood that the same definitions, procedures, and techniques that apply to fingerprints in general apply equally to all palmar and plantar impressions. No distinction is made between them from the standpoint of evidence or processing. A quality of persistence, determination, and imagination in the crime scene investigator will be most rewarding.

The first police personnel to arrive at a crime scene must be fingerprint-conscious; they must protect the scene adequately to prevent fingerprint destruction until the scene can be processed. The exclusion of people from the scene may well extend to the victim and other police officials. Everyone not absolutely necessary to the conduct of the investigation should be excluded from the immediate area of the crime scene to prevent the inadvertent destruction of evidence.

Many uninformed people are under the misapprehension that an attempt to obtain prints could be unrewarding and unsuccessful because of the physical nature of the object, or the object's having been subjected to adverse conditions. However, intelligent and persevering attempts to obtain prints should be made under all circumstances. Smudges made by fingers with a foreign substance on them, such as grease or blood, which have been placed on an object and have simply left a mark, should not be thought of as latent prints. All investigators should be able to recognize and differentiate between a smudge and a latent print, in that smudges lack ridge detail.

Latent prints fall into three general classifications: those made by a finger coated with a foreign substance such as blood, grease or dirt, which are plainly visible; those found impressed in pliable substances such as butter, candles, putty, or semidry paint; and those made by the natural body secretions, such as perspiration from the pores of the skin of the fingers, or the oils and waxes exuded by the sebaceous glands of the body, like those around the nose, ears, and neck, and picked up accidentally by the fingers. Visibility of latent fingerprints depends upon the physical condition of the person leaving them, the surface of the object, the angle of reflection of the light by which they are viewed, the time elapsing since they were made, and other factors. The amount of time they will remain on an object is also to some extent dependent upon atmospheric conditions, air currents, and humidity.

Processing the Crime Scene

Searching

It is logical to start the examination of the crime scene for fingerprint evidence at the actual point of entry. Other possible points of entry should not, however, be overlooked, since futile attempts may have been made there. The search continues in the same manner as for other evidence—that is, in a clockwise spiral around the room. A strong oblique light is a great aid to the investigator in discovering latent fingerprints. (See Figure 6.8.) Always note exactly where latent prints are found and their location on the object on which they are found. The angle they assume might indicate how an object was held, or what position the hand was in when the print was made. Sometimes, when picking up a heavy object located close to a wall, a person will place his hand on the wall as a brace. This location should not be overlooked when processing a crime scene.

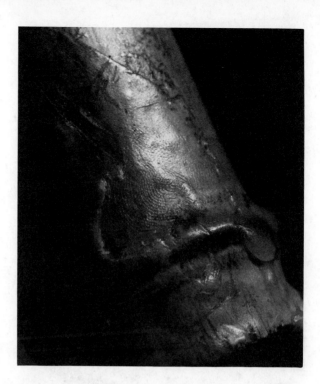

FIGURE 6.8
Fingerprint on candle

The same applies to countertops and other flat surfaces upon which a subject may lean without actually moving them. The undersides of heavy objects such as tables, chairs, and other furniture should not be overlooked as possible sources of latent prints, since it is natural for finger contact to occur when lifting or moving them.

Partial prints should be marked for orientation; for example, which is the tip end? From its location, and if other fingerprints or fingermarks are present, it may be possible to determine which finger of which hand made the latent. If two or three prints are available, it is nearly always possible to determine which fingers made them.

Numerous prints may be found at the scene, and all should be preserved; any unneeded ones can later be discarded. A notation as to exactly where and when the latent print was found, and by whom, is most important. Any slight mistake by the investigator when he testifies in court might result in the elimination of the evidence from consideration.

Photographing

When a latent print is found, the first thing that must be done, and *always* the first, is to photograph it. Various techniques may have to be tried before a photograph of value may be obtained. The use of reflected light at different angles, the use of filters, and the use of different types of photographic film may all have to be tried. The use of backlighting

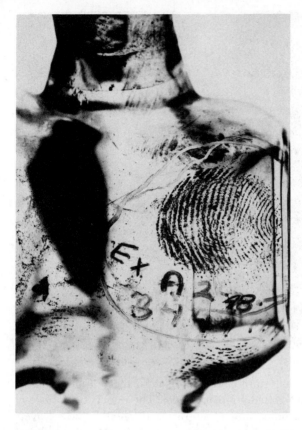

FIGURE 6.9
Backlighted fingerprint on glass tube

through a pane of glass has been very successful with the faintest of latent prints. (See Figure 6.9.)

A view camera is most versatile; by being able to view the latent print under varying conditions of light, the investigator can determine the procedures best adapted to the situation. With the lens down, good readable prints can be obtained from curved surfaces with the view camera. A camera no longer manufactured, but one that will be in use for a number of years, is the Speed Graphic. It can perform all the functions of the view camera.

Preserving

After a print has been photographed, other methods of preserving it may be attempted. Visible prints made with some foreign substance on the fingers can often be lifted with fingerprint-lifting tape. If these types of latent prints, or the plastic type, are on small objects, the entire object may be retained and held as evidence.

Latent prints made with just the normal secretions of the skin usually have to be processed in a special way before they can be of any real value. Perspiration is about 98 percent water, with the remaining percentage being composed of fatty acids, urea, sodium and potassium chlorides, phosphates, carbonates, sulphates, formic acid, acetic acid, butyric acid, and sometimes traces of albumen. These substances lend themselves to various types of treatment. The most frequently used processes are powdering and chemical treating.

Powdering

Commercial fingerprint powders have been developed over many years, until today they are dependable and of the proper composition. They come in several colors, but black, grey, and dragon's blood are the most frequently used. Choosing the powder that best contrasts with the background is a good rule of thumb to follow. Dragon's blood has the advantage of showing up on either a dark or a light background.

A test print should be made by the investigator to determine the appropriateness of the powder for the conditions present. The area selected for the test print should first be lightly brushed with powder to see if any unseen latent is actually present. Then the surface can be wiped clean and the test print made and processed. (See Figure 6.10.)

The key to successful powder development is to use a small amount with a delicate touch. A portion of the powder should be poured out of the container onto a sheet of paper, and the ends of the brush bristles just touched into the powder. Then the excess powder should be shaken off. A smooth stroke, using the fingers to guide the brush over the suspected area or over the barely visible print, is the technique to adopt. When sufficient ridge detail has been developed so that the direction of flow of the ridges can be observed, the brushing, if continued, should

FIGURE 6.10 "Powdering" a weapon

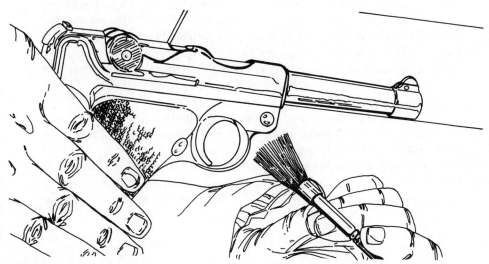

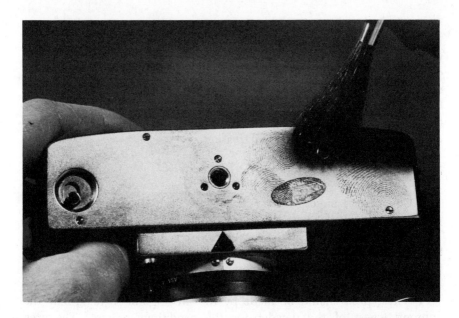

FIGURE 6.11
Developing latent prints on a camera

follow the ridge flow. When the ridge detail has been developed, it should be photographed. After it has been photographed, if it is deemed desirable, the powdering can be continued in an attempt to bring the print up into greater visibility. It may then be advisable to photograph it again. (See Figure 6.11.) The atomizer is not an effective device for developing prints under conditions usually encountered.

A new procedure, recently invented, has been adopted by numerous police departments and is used in police crime laboratories. The process combines an extremely fine metallic powder and a magnet. The unit is called a Magna Brush, although it has no bristles. The magnet picks up the powder, and only the powder touches the latent print, reducing the possibility of destroying ridge detail as a bristle brush might do. (See Figure 6.12.) Using the Magna Brush, excellent prints have been developed on wood, leather, paper, and even cleansing tissue. However, the Magna Brush process is comparatively expensive, and good results can be obtained with the normal techniques and equipment.

With ordinary fingerprint powders, a technique that has been very satisfactory for developing latents on paper, especially if the prints are fresh and the paper only semiglazed, is to allow the powder to slide back and forth over the paper without brushing. Brushing has a tendency to disturb the fibers of the paper and destroy detail; and occasionally, in spite of all precautions, the powder will adhere so tenaciously to the object on which the latent is found that brushing will not remove the excess.

Fingerprint equipment may include fluorescent powders, such as anthracine and zinc orthosilicate, for developing fingerprints on multicolored objects. A source of ultraviolet light is required. By photographing the fluorescence, the colored background is eliminated. (This method calls for some degree of skill as a photographer; therefore, it is recommended that it be used only in a laboratory.) Both a long- and a shortwave ultraviolet lamp should be available, as some fluorescent substances will fluoresce under the shortwave and others under the longwave. If the wavelength of an ultraviolet light source is unknown, sniff the air close to the light source; if ozone is detected, the light source is a shortwave. Routine use of a dragon's blood powder will produce much the same effect, without the photographic difficulties and use of ultraviolet light.

Glass objects may be passed through the flame and smoke of a burning piece of pine wood. A black, even deposit of soot will form on the

FIGURE 6.12
Dusting for latents on knife handle with Magna Brush

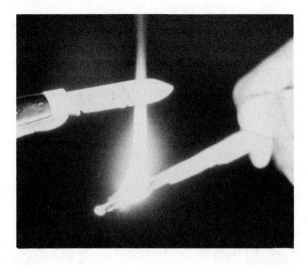

FIGURE 6.13
Soot development on knife blade

object, and careful brushing will often result in developing old latent prints. (See Figure 6.13.) This procedure may also be used to acquire a small supply of black powder when in the field without proper powders. By removing the accumulated smoke particles from time to time, sufficient powder may be obtained to process latent prints. The processing should be attempted only in the criminal investigation laboratory.

Iodine Method

The iodine method of developing latent fingerprints is not essentially a chemical one, as no chemical reaction takes place; rather, the iodine vapors are physically absorbed into the grease and oil deposits in the print. When the object is porous, like paper or unpainted wood, fuming with iodine is preferred first. This is not to say that fuming painted wood and other nonporous surfaces is precluded—sometimes it is even more effective than powder—but soft, porous surfaces lend themselves more readily to the iodine process. Iodine is very satisfactory for fresh prints not only on paper, but also on glass, finely woven fabrics, shiny-skinned fruits, cartridges, silver, and "greasy" prints.

The iodine fuming gun supplied with the fingerprint kit is ideal, but one can be easily made from any glass tube of about ½-inch diameter or a little larger. Iodine crystals are introduced into the tube and held in place by a suitable material; absorbent cotton, glass wool, and steel wool are all good for this purpose. It is advisable to introduce some calcium chloride into the end through which the breath will be blown; this can also be kept in place with any of the materials mentioned. If some such substance as calcium chloride is not used to absorb the moisture in the breath, droplets of iodine-stained water may drop from the end of the tube onto the object. The heat of the breath causes the iodine crystals to sublime faster than they would in the air, producing a usable volume of iodine vapor, which is played over the suspected area.

As soon as a print is suitably developed, it should be photographed. No attempt should be made to lift the print with lifting tape. This applies to powder-developed prints on paper as well, as the tape will tear the fibers of the paper. If it is desirable to preserve the iodine-developed print, the paper may be placed between two pieces of glass and the edges of the glass sealed with masking tape. It is also possible to lift an iodine print using a thin plate of silver. The polished silver should be held against the developed print for a few seconds and then subjected to the light from a photoflood lamp. The ridges where the iodine was in contact with the silver will turn black. The silver plate may be cleaned with whiting (a finely powdered chalk) for further use if desired.

Ammonia fumes will clear iodine-developed prints. In addition, a weak solution of "hypo," fifteen grams of sodium triosulphate to a pint of water, may also be used to clear iodine-developed prints from paper. The paper should then be washed and ironed. In this procedure, as with the silver nitrate process, to be described below, care must be taken to ensure that the paper will withstand wetting and washing. The fuming gun must be cleaned well after each use. Care must also be taken to make sure that no loose iodine crystals are in the kit and that iodine-crystal containers are tight. Iodine vapors are very corrosive and will attack metals and other component parts of the kit. A common and recommended practice is to keep the iodine crystals separate from the kit until they are needed for immediate use.

A product called "Driodine" has been developed, by grinding porous glass and saturating it with iodine, which permits actual iodine dusting without the problems normally associated with using pure iodine. Driodine is simply poured over the surface under examination and left for 15 to 30 seconds to produce an iodine-developed print. This method can be used in lieu of, or in addition to, iodine fuming.

Another method for developing latent prints on flat, porous surfaces through the application of iodine is with the use of photographic film. The film, after being completely exposed and developed, is moistened until the emulsion becomes tacky, and is then pressed against the iodine-fumed print. Then the film is dipped in a mixture of chrome alum and nitric acid, which dissolves the silver background that is not protected by the silver iodine, leaving a fingerprint that may be compared directly or used for making photographic prints or enlargements.

FIGURE 6.14
Silver nitrate process

FIGURE 6.15
Spraying with ninhydrin

Chemical Processing

All chemical processes for developing latent fingerprints are dependent upon the presence of mineral or organic matter in the perspiration. The fact that the composition of perspiration varies, not only with different people but from time to time in the same person, accounts for the uncertainty and characteristic "spotty" development of prints using chemical processes.

The most frequently used chemical process is the silver nitrate method. (See Figure 6.14.) A 3 percent solution is recommended, but it may be weaker or stronger. One and one-half teaspoons of silver nitrate to a pint of water will make about a 3 percent solution. It is advisable to use distilled water, if available, as tap water may contain undesirable chemicals. An object made of paper or cloth that is to be processed for fingerprints does not have to be soaked, just wet. After its wetting with the nitrate solution, the object should be washed to remove the excess, then placed under ultraviolet light, sunlight, or a photoflood lamp. Under ultraviolet, the images appear quickly; they appear less quickly with the photoflood.

The object itself will continue to darken with prolonged exposure, to the point where the print images are obscured. If it is desirable to clear the object, it may be placed in a 3 percent mercuric nitrate solution. After bleaching and washing, a document is placed between sheets of blotting paper and ironed with a moderately hot iron. The blotting paper is turned occasionally to dissipate the steam.

Continuous observation of the object is required during the time of light exposure, to ensure that the latent prints do not overdevelop. Latent prints developed with silver nitrate should be photographed immediately and then protected between two sheets of black paper. To preserve prints developed by the silver nitrate process, the object may be subjected to the action of ordinary photographic "hypo." The amount of contact and the amount of pressure exerted by the fingers are two of the variables that affect paper latents the most.

Old fingerprints on paper can best be developed by using ninhydrin (Figure 6.15), which is available in spray cans from fingerprint supply houses. It brings out latent prints in a dark grey or black, with good tracing. Ninhydrin should be used after the iodine-fuming method but before the silver nitrate process.

The vapors of ninhydrin are toxic in nature and care should be exercised in its use. It may also be available for use in white powder form, which must be mixed with one of several solvents, the choice of which depends upon the particular need of the user. The liquid solvent has no effect on the latent print and is used only to dissolve and dilute the chemical powder. Most solvents used may cause ink to run or smear, and for this reason, it is advisable that evidence that contains writing or typing be masked with glass plates prior to spraying.

The age of the latent prints on a document is no factor in the success of the ninhydrin process; prints have been developed that were several years old. Some prints developed with ninhydrin may appear immediately after processing; others may take several hours or even days to develop. Ninhydrin may be removed from documents by using a 1–3 percent solution of diluted ammonia.

A metal object held over a piece of burning magnesium ribbon will have deposited on its surface a film of white magnesia powder. Latents developed in this way are very persistent; even scrubbing has failed to entirely obliterate a print developed in this manner.

Lifting Fingerprints

The most common materials utilized for lifting latent fingerprints are rubber and transparent lifting tapes, which can be procured from commercial sources. Rubber lifting tape stores well and comes in black and white for use with the different-colored lifting powders. A piece of tape large enough to cover the print, while leaving plenty of room, should be utilized. The plastic cover should be carefully removed in one steady movement; any pause will result in a line being left on the tape, and in most cases, the powder on the print will not adhere to the line, thus possibly ruining the print. Apply the adhesive side of the tape to the powdered print, press it down evenly, then peel the tape from the surface. Replace the plastic cover on the tape over the lifted print to protect it.

Rubber tape is better than transparent lifting tape for taking prints from curved or uneven surfaces, but transparent lifting tape has the advantage of presenting the lifted fingerprints in the correct position, rather than reversed as on the rubber tape. Transparent tape is available in dispensers, which speed up the lifting process. The prints on transparent lifting tape should be mounted on glossy material, with the color contrasting with that of the lifting powder.

When lifting a print with either transparent or rubber lifters, care must be exercised to prevent the formation of air bubbles under the lifter. (See Figure 6.16.) Latents found in dust should first be photographed, then they may be lifted. They should not be powdered, as this will destroy them. Ordinary transparent tapes used in the home or office are not really suitable for lifting fingerprints; however, they may be used as a field expedient.

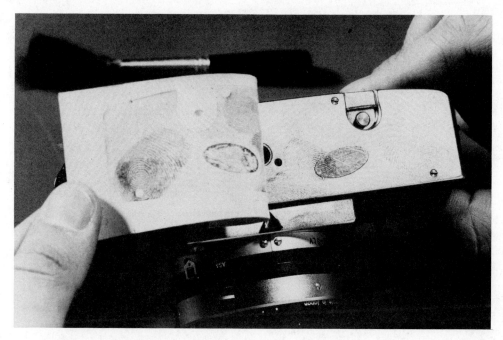

FIGURE 6.16   Lifting the developed prints (see Figure 6.11)

FINGERPRINT PATTERNS

Every investigator should be sufficiently familiar with the general pattern types so that he can at least differentiate between them. If an investigator has a working knowledge of the classification procedure, he will obtain more intelligent fingerprint evidence at a crime scene, as he will know what to look for and will not be misled by smudges or other fingerprints.

Classification Designations

Fingerprint patterns have three basic classification designations: arches, loops, and whorls.

Arches

The classifier divides *arches* into two types: plain and tented.

Plain arches are those types of patterns in which the ridges enter on one side of the impression and flow, or tend to flow, to the other with a rise or wave in the center, making no upward thrust, backward looping turn, or angle, and possessing not more than one of the three basic characteristics of the loop. (See Figure 6.17.)

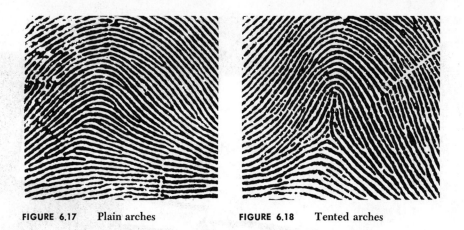

FIGURE 6.17   Plain arches   FIGURE 6.18   Tented arches

Tented arches are similar to plain arches except that the ridges in the center form a definite angle, or one or more ridges at the center form an upthrust or approach the loop type, possessing two of the basic or essential characteristics of the loop but lacking the third. (See Figure 6.18.)

Loops

A *loop* is that type of fingerprint pattern in which one or more of the ridges enter on either side of the impression, recurve, touch or pass an imaginary line drawn from the delta to the core, and terminate, or tend to terminate, on or toward the same side of the impression from which they entered. The loop must have three essential characteristics: a sufficient recurve and its continuance on the delta side until the imaginary line is reached; a delta; and a ridge count of at least one.

The classifier divides loops into two types: ulnar and radial. *Ulnar*

FIGURE 6.19   Ulnar loop, right hand   FIGURE 6.20   Radial loop, right hand

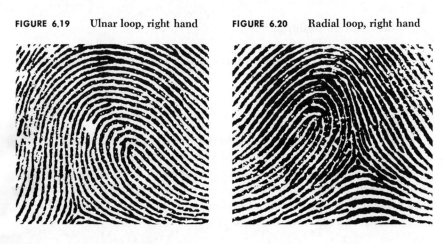

loops are those types of patterns in which the loops flow in the direction of the little fingers (Figure 6.19); *radial loops* are those types in which the loops flow toward the thumbs (Figure 6.20).

Whorls

Whorls are those types of patterns in which the ridges form concentric circles or spirals, or some variant of this geometric form. They are divided into plain whorls, central pocket loops, double loops, and accidentals.

A *plain whorl* (Figure 6.21) has two deltas and at least one ridge making a complete circuit, which may be spiral, oval, or any variant of the circle. An imaginary line drawn between the two deltas must touch or cross at least one of the recurring ridges within the pattern area.

The *central pocket loop* (Figure 6.22) consists of one or more re-

FIGURE 6.21 Plain whorl (top)   **FIGURE 6.22** Central pocket loop (top)

FIGURE 6.23 Double loop (bottom) **FIGURE 6.24** Accidental (bottom)

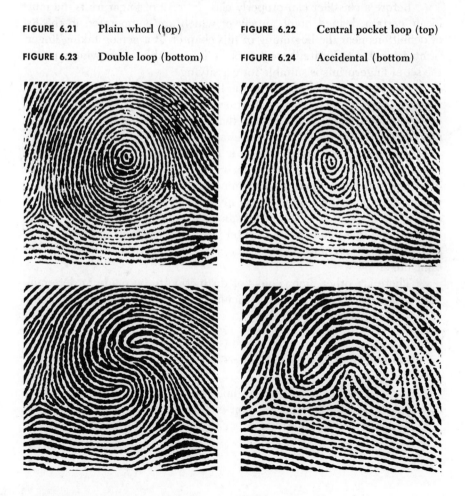

curving ridges, or an obstruction at right angles to the inner line of flow, with two deltas between which an imaginary line would cut or touch no recurving ridge within the pattern area. The inner line of flow of a central pocket loop is determined by drawing an imaginary line between the inner delta and the center of the innermost recurve of looping ridge.

The *double loop* (Figure 6.23) consists of two separate loop formations, with two separate and distinct sets of shoulders, and two deltas.

The *accidental whorl* (Figure 6.24) is a pattern with two or more deltas, and a combination of two or more different types of patterns exclusive of the plain arch. This classification also includes those exceedingly unusual patterns that may not be placed by definition into any other class.

Interpretation

Before a classifier can properly classify a set of fingerprints, he must locate certain defined points, many of which are enumerated in the list of definitions near the beginning of this chapter. If a person taking fingerprints is unable to locate these points, he is incapable of knowing whether the set of fingerprints is suitable for classifying.

For instance, a plain arch usually does not have a core or delta, whereas a tented arch may have either or both. A loop has one core and one delta. A whorl may have more than one core, must have two deltas, and frequently has more than two deltas.

In order to classify a loop, it is necessary to count the number of ridges that cross or touch an imaginary line drawn from the delta to the core, and this has a bearing on elimination procedures. It should be apparent that, if a latent print differs in pattern from an elimination print, or if a latent print is a loop with about 15 ridges intervening between the core and delta and an elimination print is also a loop but with only 7 or 8 ridges intervening, both prints could not have been made by the same finger.

There are other, finer points of distinction to be made in determining exactly what constitutes a particular pattern, but it is not our intent to cover them here.

Classification and Identification

Many fingerprints are similar, and many fingerprint cards will have the same classification. Classification is simply a method by which a set of fingerprints may be suitably filed in a filing cabinet and easily retrieved

Fingerprinting 145

for future use. Just as books in a library and correspondence papers are given filing numbers so that those dealing with the same topics are filed together, so are fingerprint cards.

Identification in fingerprinting is the determination that in two fingerprints, there are so many identical characteristics alike in structure, appearing in the same relative places, that they are the same and could not have been made by two different fingers but only by the same finger. (See Figures 6.25 and 6.26.)

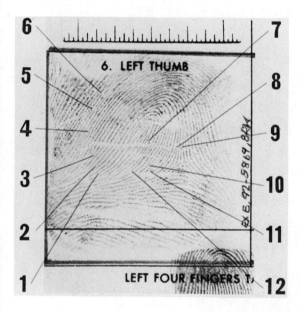

FIGURE 6.25
Print obtained from fingerprint card (record)

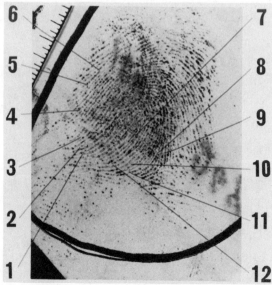

FIGURE 6.26
Print discovered at crime scene (latent)

MAINTENANCE OF FINGERPRINT RECORDS

Because fingerprint records are essential in the identification of people, the data provided must be maintained in a manner that will not result in their compromise or accidental destruction.

The investigator records all data concerning fingerprints in the notes of the investigation, and the location where fingerprints were found at a crime scene are cross-referenced to the crime scene sketch as well. This assists him in the conduct of the investigation and in his later preparation for testimony in court, should that become necessary.

7

Firearms Identification

INTRODUCTION

The successful solution of a crime involving firearms may depend entirely upon the collection and preservation of firearms evidence by a knowledgeable investigator and detailed examination of the evidence by an expert technician at the police criminal investigation laboratory.

DEFINITIONS

Firearms identification: The technical examination of arms and ammunition by a qualified laboratory technician.

Firearms: All small arms and shoulder or hand weapons, including all types of rifles, pistols, shotguns, and submachine guns, as well as various light machine guns.

Ammunition: Cartridges that can be fired by the firearms defined above, and the component parts of such cartridges—the bullet, cartridge case, primer, and propellant (powder).

Shotgun: A shoulder arm having a barrel that is smooth-bored and is intended for the firing of a charge composed of one or more round balls or pellets of shot.

Pistol: A hand arm having a short barrel, usually not over 10″ long, with a grip designed so that it can be held in and fired from one hand with the arm extended. Although revolvers are often included in the general classification of pistols, we prefer here to define a pistol as a hand arm in which the cartridge is loaded directly into the chamber of the barrel, to distinguish it from a revolver, in which the cartridge is loaded into a cylinder in the rear of the breech of the barrel.

Single-shot: Refers to a gun, rifle, or pistol that loads with only one cartridge. To fire it a second time, the breech mechanism or action must be opened, the fired case or shell extracted, and another cartridge inserted in the chamber.

Double-barrel: A small arm having two barrels, placed either side by side or superimposed, each having its own firing mechanism.

Repeating: Refers to a small arm that has a magazine or recess in the breech action or under the barrel or stock, and that, after firing, requires hand action of the operator to eject the fired case and chamber a new cartridge.

Semiautomatic: A weapon operated automatically by the ammunition it uses, but requiring release and re-pressing of the trigger for each succeeding shot.

Automatic: A repeating firearm that utilizes the power of the recoil or a portion of the expanding gases to complete the cycle of ejecting the fired case and chambering another cartridge. An automatic arm will continue to fire with rapidity as long as the trigger is depressed or until the magazine is empty.

Carbine: A short-barreled musket or rifle, having a barrel length generally not longer than 22″.

Caliber: The diameter of the bore of rifled small arms. In the United States and all English-speaking countries, it is designated in hundredths or thousandths of an inch. Thus we have rifles, carbines, pistols, and revolvers of .22, .25, .30, and .45 caliber, or .220, .257, .357, and .405 caliber. On the continent of Europe and elsewhere, the caliber of such weapons is designated in millimeters, as 6.5 mm, 7 mm, 8 mm, or 9 mm. The designation, for example, of 7.65 mm is .32 caliber, of 6/35 mm is .25 caliber, and of 9 mm is .38 caliber.

Extractor: That mechanism in a firearm by which a cartridge or fired case is withdrawn from the chamber.

Ejector: That mechanism in a firearm that throws the cartridge or fired case from the firearm.

RESPONSIBILITIES

The investigator is responsible for collecting, preserving, and forwarding firearms evidence to the laboratory for examination. He should be aware of what information the laboratory can provide and how this can assist him in his investigation. The investigator does not perform firearms identification tests in the field unless he is thoroughly qualified.

The laboratory receives firearms evidence, conducts firearms identification tests, and furnishes the test results to the investigator. The laboratory technician renders expert testimony in a court of law when required to do so.

Firearms Information

The main problem is to establish whether or not a fired projectile or expended cartridge case submitted as evidence was fired in or from a specific weapon. However, additional information may be developed by the laboratory technician that may be of great value in achieving a successful conclusion to an investigation. Such additional information may include the following:

1. The mechanical condition of a weapon may provide indications that an accidental discharge was possible. Tests may disclose whether or not the weapon was fired since it was last cleaned.

2. A fired bullet or cartridge case may reveal the caliber and make of the weapon that fired it.

3. The manufacturer of the ammunition (commercial, military, foreign, etc.) may be determined.

4. The distance between the muzzle of the weapon and the point of contact may be adequately determined.

5. The point of entrance and/or exit of a projectile in clothing, wood, glass, metal, and the like, may be established.

Modern police crime laboratory techniques are highly scientific and require the use of a variety of sophisticated equipment. Testimony in this field should be presented only by a qualified technician, testifying as an expert witness. The technician must be able to qualify before a court of law to give expert testimony in the field of firearms identification.

In view of more definite court rulings, the successful solution and prosecution of crimes are increasingly more dependent—in some instances, entirely dependent—on the evidence developed by the laboratory technician. It is essential, therefore, that the investigator understand how the firearms examiner can assist him, and must also be fully aware of his own responsibilities in the proper collection and preservation of firearms evidence in the conduct of his investigation.

EVIDENCE

Recovery of Evidence

The investigator may experience difficulty in recovering fired bullets at a crime scene, since they may be lodged in ceilings, walls, furniture, or flooring. When removing a bullet from its resting place, he should be careful not to mutilate any of its identifiable features. Sometimes, it is desirable to remove a small section of the wall or ceiling and forward it intact to the laboratory rather than damage the bullet. Accurate notes are recorded as to the location and condition of the bullet, the type of material it pierced, the depth of penetration, irregularities of size and shape, approximate angle of impact, and other pertinent information that may assist the laboratory technician in the examination. The point at which each discharged bullet or fired cartridge case was discovered is carefully and accurately recorded in the investigator's crime scene sketch.

When a bullet is lodged in a body, the investigator advises the surgeon of its potential value as evidence, and requests that he exercise extreme care in probing, so that its surface will not be damaged by the probe. Rubber-tipped forceps should be used to remove the bullet, and after recovering it, the surgeon should place some identifying marks upon it so that he can identify it later. The investigator should inform the surgeon as to the proper method of marking the bullet.

In many instances, the surgeon is required to cut away the clothing of the victim of a gunshot wound when rendering medical aid. If at all possible, the investigator should take steps to ensure that cuts in the clothing material do not seriously alter the entrance or exit bullet holes or areas that might contain gunpowder residues. Clothing removed from the victim should also be marked for identification at the time of removal, or as soon thereafter as practical.

In cases involving shotguns, a thorough search is made to recover all wadding or shot columns. These items of evidence are most important in determining the manufacturer of the ammunition used. The shot column may sometimes be used to positively identify the shotgun from which it was discharged. The presence of wadding or shot column inside the body of a shooting victim indicates that the firearm was at very short range.

It may be that the perpetrator of a homicide who uses a firearm will attempt to conceal the crime by setting fire to the premises, thereby destroying the evidence. In instances of apparent death by burning, it is advisable to request that the remains be X-rayed. This procedure may reveal the presence in the body of one or more bullets and label as murder or suicide what previously appeared to be accidental death.

Marking of Evidence

Evidence is properly marked in order that it may be readily identified in the future. Firearms determined to be of evidentiary value are marked for identification; however, firearms confiscated or impounded for later determination of evidentiary value are not to be marked, scratched, or defaced in any way until a determination is made that they have evidentiary value.

Firearms evidence will normally comprise various types of exhibits, which should be marked in certain ways. The investigator places his initials, the date of recovery, and the time on each item of evidence so that he can positively identify it at a later date. Where several items of identical or similar appearance are recovered, he should add an identifying letter or number on each item, so that no two items in the same case bear exactly the same markings. All identifying markings and a description of items to which they are affixed should be recorded in the investigator's notes. The identifying letter or number, if such is used, will have no bearing on or relation to the letter designation of the exhibit in the report of investigation. (For example: Three bullets recovered during the course of an autopsy of the victim—one from the chest, one from the left leg, and one from the right upper arm—should be marked with a letter or numerical designation in addition to date and initials. At a future date, a differentiation can be made as to which particular projectile was recovered from which area of the body.)

A variety of marking tools may be used for the purpose of inscribing identifying markings on firearms evidence. Diamond-point or Carborundum pencils are ideal for this. Dental picks, available at dental clinics and dentists' offices, also make an excellent marking device, when the curved tip is cut off the tool and the point sharpened to a needle-sharp tip. Lacking these tools, and as a last resort, the point of a knife blade, the corner of a file, or a corner of a small Carborundum stone may be used for marking.

Firearms are usually marked on the right side of the frame, with any suitable marking tool. All removable parts of the firearm that leave imprints on either the bullet or cartridge case are marked. For example, a conventional .45 semiautomatic pistol should be marked on the barrel (which marks the bullet), the slide (which contains the firing pin, extractor, and breech face, each of which marks the cartridge case), and the receiver (which includes the ejector, which also marks the cartridge case). All these parts should be marked alike. (See Figure 7.1.) The markings are placed on the slide, the barrel, and the frame. All portions are marked in an area where the marking can be seen readily and does not interfere with existing markings, stampings, or parts of the weapon. The magazine is marked and submitted with the suspect weapon.

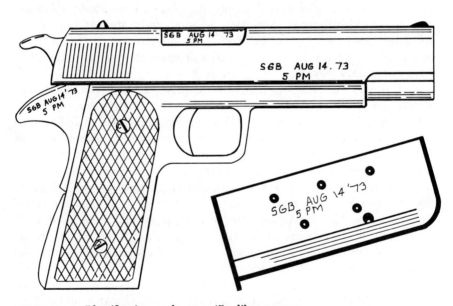

FIGURE 7.1 Identification marks on a .45-caliber weapon

Revolvers are marked on the frame and cylinder. The mark goes on the side of the frame, which cannot be removed. (See Figure 7.2.)

Weapons having a movable bolt (such as a bolt-action weapon, or the semiautomatic and automatic weapons having a movable bolt) are marked on the bolt, barrel, and frame. If the barrel of a weapon cannot

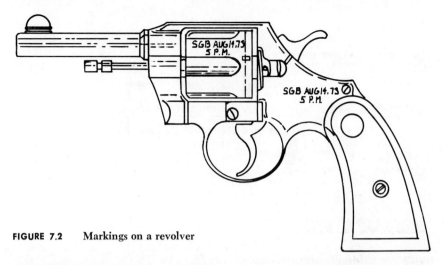

FIGURE 7.2 Markings on a revolver

be removed without tools, it is not necessary to mark it, but even under these circumstances, doing so is added insurance for the investigator.

A fired bullet submitted as an exhibit may be either jacketed, as is military ammunition, or lead, which is common with commercially manufactured revolver ammunition. Markings may be placed on the base of the bullet, if sufficient area exists and such marking will not obliterate the manufacturer's marking or other marking that may be in this location. If the base cannot be marked, the nose is marked, beyond the upper limit of the rifling imprints on the cylindrical surface of the bullet. (See Figure 7.3.) A bullet is never marked in such a manner that the rifling imprints will be defaced.

Deformed bullets and jacket fragments may be marked wherever practical—that is, on any exposed area that is free of rifling imprints and is large enough to receive identifying markings. Jacket fragments bearing rifling marks on one side may be marked on the smooth (reverse) side. Core fragments of sufficient size may be marked at any convenient point.

FIGURE 7.3 Identification markings on bullets

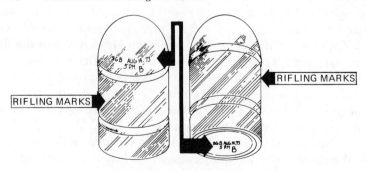

CHOICE OF LOCATION

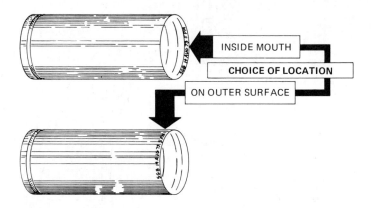

FIGURE 7.4 Identification markings on cartridge cases

If the bullet fragments are too small to be marked, they are treated in the manner described below for shot pellets.

Large-caliber cartridge cases are marked within the mouth, and small-caliber cartridge cases are placed in a container that is marked. They are never marked on the head (base). (See Figure 7.4.) Chambering marks running in line with the longitudinal axis of the cartridge case may be etched on fired cartridge cases. These marks may be of value to the laboratory technician and should not be marked over. They are difficult to see, however, and a careful search in good lighting is often necessary to locate them. Markings are always made on clean areas, free of blood, dirt, or other adherent substances that could be significant to the laboratory technician.

Shotgun shell cases may be of brass, heavy paper, or plastic material. When they are made of paper, they may be marked in ink. When brass, they may be marked in a manner similar to that used for cartridge cases. When the case is plastic (which is becoming most prevalent), it is normally better to mark on the metal collar just below where it joins the plastic cylindrical body of the shell case. Shotgun wads of felt or composition material cannot be marked and are placed in separate containers and labeled. Paper wads may be marked in ink, provided this does not deface printing appearing upon them or the marks on the undersurfaces of "overshot" wads where individual shots have been in contact with such surface. To avoid defacing such printing or shot marks, the same procedure should be followed as in the case of felt and composition wads.

Shot pellets (birdshot, buckshot, and so forth) do not lend themselves to individual marking. All shot known to be from the same source is placed together in a suitable container and the container sealed and properly marked for identification.

Plastic shot columns, common in many shotgun shells, are projected through the barrel of the shotgun and sometimes bear markings similar to rifling imprints. These columns are marked on the base. (See Figure 7.5.)

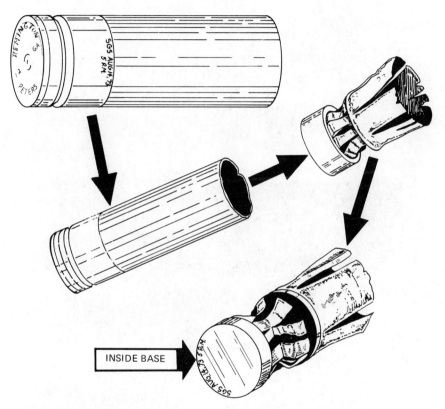

FIGURE 7.5   Identification markings on shot shells and shot columns

Transmittal of Evidence

Physical evidence that may require examination by a firearm technician is handled with the utmost care to eliminate any possibility of alteration or destruction. Contamination is also a serious consideration, particularly when clothing and like items are involved. Special care is exercised in packaging such evidence for transmission to the criminal investigation laboratory.

Firearms to be examined at the police crime laboratory are unloaded prior to being shipped. They are wrapped in a clean protective covering

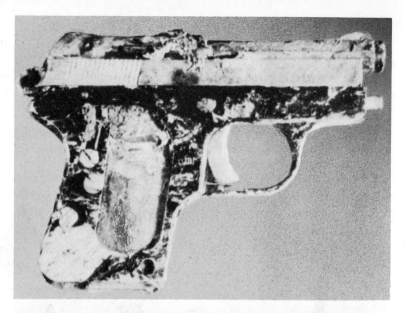

FIGURE 7.6   Weapon located in shallow pond (top)

FIGURE 7.7   Same weapon, restored by laboratory technician (bottom)

to prevent dust, lint, and other foreign matter from filtering into the mechanism, and then packed in substantial shipping containers. When the evidence is to be examined for fingerprints, special packaging is required and annotated appropriately.

Under normal conditions, firearms are not cleaned before shipping. The only reasons for the investigator to clean the weapon are if ammunition with a corrosive primer was fired through the weapon, or if there is considerable moisture inside the barrel of the weapon. Corrosion and rust make it difficult, but not impossible, for the technician to conduct comparison tests. For example, a positive identification was established where a revolver had been buried in damp soil for a period of more than three months. The exterior of the weapon was a solid mass of rust, but even though rusting had also taken place in the interior of the barrel, it was still possible to make an identification. (See Figures 7.6 and 7.7.) The nearest police crime laboratory should be consulted in special cases where firearms must be cleaned. Firearms found submerged in water should be shipped in the liquid in which they are found, or submerged in oil in leakproof containers.

Any ammunition discovered in the possession of a suspect or at the scene of a crime is seized and held as evidence. (See Figure 7.8.) The laboratory normally maintains sufficient ammunition of a similar type to use for test purposes. In the event that the laboratory technician has difficulty in finding the right type of ammunition, he may ask for samples from the seized material. In those instances in which evidence is hand-carried to the laboratory, sample ammunition taken from the suspect or the crime scene should be taken along for possible test purposes.

FIGURE 7.8 Ammunition for comparison

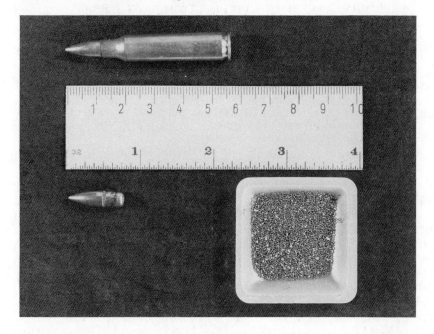

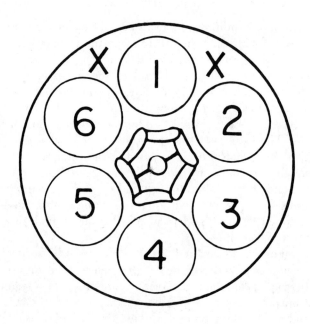

REAR FACE OF CYLINDER

1. NUMBER CHAMBERS CLOCKWISE, 1 TO 6.
2. LIST CHAMBER CONTENTS IN THE FOLLOWING MANNER:

   A. **CHAMBER 1.** DISCHARGED CASE (STEEL), HEADSTAMPED "ECS 43." MARKED FOR IDENTIFICATION "SGB AUG. 14 '73 5 PM B"
   B. **CHAMBER 2.** EMPTY.
   C. **CHAMBER 3.** BALL CTG. (BRASS CASE), PRIMER INDENTED (MISFIRE), HEADSTAMPED "FA 45." MARKED FOR IDENTIFICATION "SGB AUG. 14 '73 5 PM C"
   D. **CHAMBER 4.** BALL CTG. (BRASS CASE), HEADSTAMPED "WRA 45." MARKED FOR IDENTIFICATION "SGB AUG. 14 '73 5 PM D"
   E. **CHAMBER 5.** EMPTY.
   F. **CHAMBER 6.** DISCHARGED CASE (STEEL), HEADSTAMPED "ECS 42." MARKED FOR IDENTIFICATION "SGB AUG. 14 '73 5 PM E"
3. X MARKS ON EITHER SIDE OF CHAMBER NO. 1 BRACKET THIS CHAMBER AS THAT IN LINE WITH BORE (I.E., UNDER THE FIRING PIN) WHEN REVOLVER WAS FOUND.

CASE NO: HOMICIDE 1713-AJ-16-L   DATE: AUGUST 14, 1973
WEAPON: 45 S&W REVOLVER M1917   SER. NO: 267-36-2490
SUBMITTED BY: DET S.G. BAXTER

FIGURE 7.9   Sample cylinder diagram and legend

When revolvers containing either loaded cartridges or fixed cases are recovered, a diagram is made of the rear face of each cylinder, showing the position of the loaded cartridges or the fired cases with respect to one another, and to the firing pin. This diagram, complete with legend, is detailed to enable the laboratory technician to associate the fired cartridges with the chambers of the cylinder from which they were fired. The reason for this is that all cylinder chambers may not line up with the chamber end of the barrel in the same manner, nor will the fired cartridge cases be marked identically, owing to a slight repositioning. The investigator can simplify this reconstruction by scratching an X on the rear face of the cylinder on each side of the chamber lying under the firing pin when the revolver was found. (See Figure 7.9.)

Discharged bullets and cartridge cases intended for transmittal to a crime laboratory should be wrapped in soft tissue or cotton to protect them from abrasion or other damage, and then packed into a pillbox or similar suitable container. The container is labeled to identify the contents with the case under investigation.

Items of wearing apparel submitted to the firearms division for proximity tests are packaged so that the area surrounding the entrance hole is protected from contamination. This is best accomplished by sandwiching the area of the garment containing the gunshot residues between sheets of cardboard. Bloodstained clothing is air-dried prior to packaging. Medical personnel should be cautioned against cutting through bullet holes when removing clothing from shooting victims.

Firearms may be shipped through the U.S. mails in accordance with current postal laws and regulations, available at any post office. However, under no circumstances may live ammunition, propellant powders, primers, or explosives be sent through the U.S. mails. Such items are shipped by railroad or air express, or transported by a courier.

Fingerprints on Firearms

Situations may arise where it is felt that a weapon should be processed for latent fingerprint impressions. In handling a firearm, the investigator exercises great care not to obliterate possible latent impressions unless he is certain that they will not be material to the case. In most cases, weapons may be picked up by the grips, since the checkering precludes obtaining usable prints from this area, or by the use of a piece of wire or some similar material inserted through the trigger guard, lanyard ring, or similar area. Do not use a handkerchief or like material or insert any object into the bore to pick up the weapon. Care should be exercised in transmitting the weapon to the laboratory, including proper packaging methods, to ensure that nothing comes in contact with the suspected surfaces that may obliterate the impressions and possibly void an essential lead in the investigation.

LABORATORY EXAMINATION

Proximity Tests

In gunshot wounds, powder residues may be deposited either on skin or on clothing. Proximity tests are based on the dispersion of these gunpowder residues. By firing the same weapon, using the same type of ammunition, tests can be conducted and the approximate distance between muzzle and point of contact established. (See Figure 7.10.) This method, of course, is subject to limitations. Normally, with a muzzle contact distance in excess of about two feet, no discernible gunpowder residues will be noted, as they are very light in weight, and will not carry any great distance, owing to air resistance and other factors.

Scaled photographs, either black-and-white or color, of gunshot wounds on flesh permit the rendering of only a qualified statement concerning proximity. Such observations as that the wound or wounds were inflicted at or near contact, or at less than or greater than 24 inches, for instance, will probably be noted, but more exact opinions will normally be precluded. Shot dispersion in shotgun shootings may result in a determination by the laboratory of proximity (again, the same type of ammunition and same weapon must be used for test purposes).

Restoring Serial Numbers and Other Markings

In many cases, firearms from which the serial numbers or other stamped lettering have been removed by grinding, filing, or some other means are discovered at a crime scene or during the course of an investigation. As with other such items—cameras, watches, and so on—establishing ownership of such property or otherwise identifying it may hinge completely on the serial number. Crime laboratory personnel can be of material assistance in this respect. The investigator should not make any attempt on his own to restore such numbers or markings.

Class Characteristics

In the absence of a suspect firearm, it is possible, in most instances, to determine by examination of a fired bullet or a cartridge case alone certain facts concerning the class characteristics of the firearm involved.

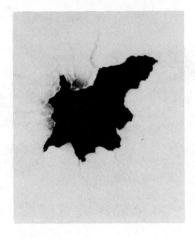
AT CONTACT

AT 2 INCHES

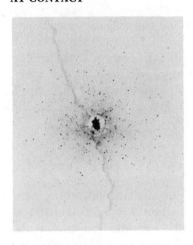
AT 4 INCHES

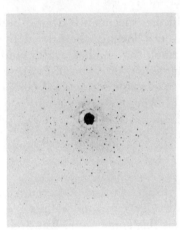
AT 8 INCHES

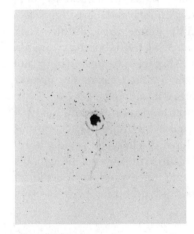
AT 10 INCHES

AT 14 INCHES

FIGURE 7.10 PROXIMITY TESTS USING .45 AUTOMATIC, FIRED INTO BLOTTING PAPER

FIGURE 7.11   Barrel rifling

If the evidence specimen is a fired bullet, the class characteristics may comprise the following: the caliber, the type of firearm (such as pistol, revolver, rifle), the number, width, and depth of grooves in the rifling, and the angle and direction of twist. (See Figure 7.11.) From this information, it is possible for a crime laboratory to provide a list of firearms capable of having fired the submitted bullet. Similar information may be obtained from the examination of a fired cartridge case.

Test Firing

If a suspect firearm is submitted in conjunction with either fired bullets, cartridge cases, or both, an examination will be conducted to determine if this ammunition was fired by the submitted weapon. If the class characteristics of the expended ammunition are consistent with the ammunition fired from a weapon of the same type as the exhibit weapon, test firing must be conducted and the test bullets and cartridge cases microscopically compared with the exhibit items. (See Figures 7.12 through 7.15.) All test firing will be conducted at the laboratory. Should a large quantity of firearms be involved as suspect weapons, it may not be feasible to ship them all for examination. In such cases, the investigators should contact the supporting laboratory for advice.

Often overlooked, but of utmost importance, is determining if a firearm will function and if it functions safely. Oftentimes, it can be established as impossible for a firearm to have discharged "accidentally" as stated by a suspect. On the other hand, a firearm may not be capable of firing at all. It is suggested that all firearms uncovered during the investigation of homicides, suicides, assaults, and robberies be submitted for function testing.

FIGURE 7.12 Laboratory-fired bullet

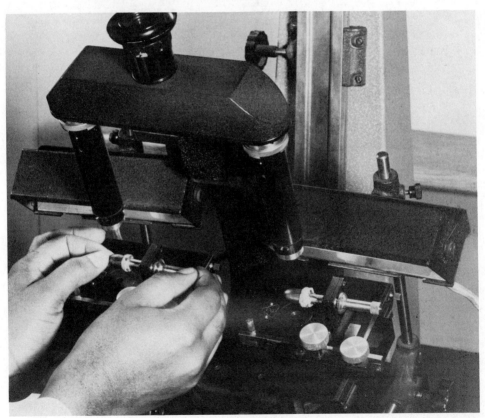

FIGURE 7.13 Positioning fired bullet for microscopic comparison

FIGURE 7.14   Microscopic comparison of striations

FIGURE 7.15   Microscopic comparison of cartridge cases

Neutron Activation Analysis (NAA)

The process of Neutron Activation Analysis, made available through the FBI, may assist in determining whether or not a person fired a certain firearm. The investigator using this process must prepare a paraffin lift of both the subject's hands, as described below, for submission to the

FBI laboratory in Washington, D.C. Prior to sending the evidence to the FBI laboratory, telephone coordination should be made with the nearest police crime laboratory.

Material needed to prepare paraffin lifts for analysis is as follows:

1. Paraffin
2. Glass beaker
3. Heat
4. Rubber gloves
5. Brush

Wearing rubber gloves throughout the process, place the beaker over heat, add the paraffin, and heat till it melts. Test its temperature by using the end of the brush to place a small amount on the inside of your forearm; take care that it is not so hot that the subject is burned.

Brush paraffin on the index finger, web, and thumb of the subject, not less than ⅛ inch thick. Allow it to dry, and remove it. Repeat with the subject's other hand.

Submit to the laboratory both paraffin lifts, a control portion of paraffin, the rubber gloves, and the brush.

8

Glass Fractures

INTRODUCTION

This chapter introduces the subject of glass fractures, as it pertains to the needs of the investigator and the laboratory analyst in an investigation wherein glass evidence is involved. It discusses the nature of glass, types and causative agents of glass fractures, stress lines, and glass fracture and fragment examination by the investigator and by the scientific laboratory.

EXAMINATION OF GLASS EVIDENCE

Glass, or glass fragments, are frequently important factors in the investigation of offenses such as burglary, arson, "fleeing-the-scene" vehicle accidents, shootings, and others. The value of such material, either as evidence in itself or in the investigative leads that may be obtained from it, is dependent in great measure on the knowledge and training of the investigator in the nature of glass, proper procedures of collection, preservation, and examination, and what he and the scientific laboratory technician can learn from it.

Glass fractures and glass fragments can, with the use of relatively easy and rapid examination techniques, provide information from which determinations such as the following may be possible:

1. That a fragment of glass did or did not originate from a particular glass object that has been broken—for example, that a fragment of headlight lens found at the scene of a vehicle accident did or did not originate from a broken headlight of a suspect vehicle

2. That a fragment of glass originated from a particular kind of glass object, such as a headlight lens, spectacle lens, window pane, etc.

3. The origin and direction of a fracture—i.e., what caused it and the direction from which the causative force came

4. In the event of multiple fractures, including bullet holes, the order in which the fractures occurred

5. The angle from which a bullet struck a glass object

6. That a particular glass object, such as a bottle or jar, contained an inflammable or explosive substance—e.g., in an arson or sabotage case

THE NATURE AND PROPERTIES OF GLASS

Glass is normally a fused mixture of silica, usually in the form of natural sand, and two or more alkaline bases, such as soda, lime, or potash. It also contains quantities of various other elements and metals, present either as incidental impurities in the basic constituents or added to them for color, degree of hardness, heat resistance, and other specific purposes. The constituents are melted in a crucible at a very high temperature, and the molten mass is then either rolled, blown, or molded into desired sizes and shapes. It may later be polished, ground, or cut for utilitarian or decorative purposes, or it may be combined with other materials. For example, sheet vinyl plastic is fused between sheets of plain glass to form safety glass, as found in cars, trucks, trains, aircraft, and so on.

The importance of glass to the investigator, either as evidence or in the development of investigative leads, lies mainly in its physical properties. These properties make it possible to determine that glass fragments did or did not originate from the same source, or to determine the manner in which a piece of glass was broken. For example, differences between two batches of glass in the amounts of mineral ingredients used to make them will produce variations that can be detected and proved under laboratory procedures. Other characteristics of glass—such as the fact that it seldom breaks squarely across but leaves convex-concave edges, or stress lines, on the fractured edges; that it bends and stretches before breaking; and that breaks produce both radial (primary) and concentric (secondary) fractures—can enable other determinations to be made by the investigator.

COLLECTION AND HANDLING

Methods used in the search for and collection of glass and glass fragments are usually the same as those used for other types of evidence. Differences are due to the nature of glass itself, to the investigative and evidentiary finds sought from it, and to its value to the investigation.

Records

Like other types of physical evidence, glass and glass fragments must be photographed and their locations noted on the crime sketch before they are touched or moved. Pertinent data must be recorded in the investigator's notes concerning the glass and any obvious, suspected, or hypothetical relation it has to the incident under investigation.

Collection

In collecting glass or glass fragments, the investigator carefully avoids smudging any fingerprints or disturbing any substance such as dust or dirt, bloodstains, or other foreign matter that may be on the glass, since any or all of these may provide investigative leads or be evidence in themselves. Rubber gloves should be worn, and rubber-tipped tweezers or a similar device should be used for handling small fragments, so as not to scratch the glass. Metal tweezers with adhesive tape placed over the inner surface of the points make a good field expedient. Glass should be picked up by the edges, and the plane surface avoided. All available fragments should be collected, to provide as complete a reassembly as possible. Even particles too small to permit matching or reconstruction should be collected and preserved, since they can be analyzed in the laboratory for their physical properties.

In those instances where glass has been broken out of a window, door, or similar frame, and pieces remain therein, the frame should be removed and kept intact, if at all possible. This method will also facilitate reassembly of the broken pieces. If such removal is not possible, the pieces remaining in the frame should be carefully marked (inside and outside surfaces must be designated) and removed in such manner as to avoid further damage to the glass or disturbance of any deposit or substance on it. If the frame is not removed, samples of the wood, paint, putty, and any other materials should be collected from it.

Shattered or broken glass may possibly provide a clue in the identification of a suspect. Often, in the commission of a crime such as burglary, the perpetrator will break a window or other glass object, and particles or fragments of the glass may lodge in or adhere to his clothing or fall into a pocket or trouser cuff. Collection and laboratory analysis of such fragments for comparison with glass found at the crime scene is a worthwhile effort. The soles and sewn edges of shoes should not be overlooked in the search for this type of evidence.

A thorough search may also be productive in the investigation of a hit-and-run vehicle accident where glass was broken. Fragments or particles of glass may be found adhering to or imbedded in the tires of the suspect vehicle. These should also be forwarded to the police crime laboratory for analysis and evaluation.

Marking of Glass

Fragments of sufficient size should be marked with a diamond-point or Carborundum pencil, a piece of properly marked adhesive tape, or a grease pencil; markings are placed in an area where there is no deposit of value as evidence. Markings should include the investigator's initials, the date, and the time. To aid in reassembly of the fragments and in reconstruction of the incident, markings should be placed on the side that was up when found, and should include a sequence number that, when keyed with the investigator's notes, photographs, and sketches, serves to identify the location where found. Fragments too small to permit such markings are placed in suitable containers and the containers marked in ink for submission to the laboratory.

Preservation of Glass

Glass fragments should be wrapped in soft paper, cotton, or similar material that will prevent breakage, in a manner that will prevent damage to fingerprints preserved as evidence. The wrapped glass is placed in suitable containers, properly fastened so that it will not shift. Wrappings and containers should be marked "Fragile," in letters large enough to be readily seen.

Methods for the safeguarding of evidence apply to glass just as to other evidence. Property tags and receipts are prepared by the investigator, and the property is released to the designated evidence custodian at the earliest practicable time. A request for laboratory examination is prepared by the investigator, who obtains release of the property from the evidence custodian for transmittal to the laboratory.

EVALUATION OF FRAGMENTS AND FRACTURES

The investigator evaluates glass fragments and fractures as he does other items of evidence, considering their value and importance both in themselves and in their relationship to all other evidence.

The evaluation begins in the initial stages and continues until the investigation is completed. In the evaluation process, the value of certain materials found on the scene is not always readily determinable. This raises the question for the investigator as to whether he should collect and preserve such materials. The safest decision, if there is any doubt, is to do so; if he does not, it is likely that the broken glass will be quickly discarded as trash, and neither the investigator nor the laboratory will have any later opportunity for evaluation or examination. Material that proves valueless can be disposed of at a later time, either during or after the investigation.

The investigator should also consider the need or desirability of scientific laboratory analysis. Should he decide upon such action, it is usually best to request it as early as possible; sometimes speed may be essential, as in cases involving fragments of glass containers that are suspected of having held inflammables or explosives, to prevent loss of odors or residue or to provide further investigative leads or clues.

General Examination

Fractured glass found at the scene of a crime may reveal the direction from which a blow was struck—for instance, if the fracture was caused by a bullet, the direction from which the bullet entered the glass and the angle from which it was fired. When broken glass is reconstructed, some of the fractures (the primaries) in the glass will resemble the spokes of a wheel, in that they will radiate outward from the point of impact. These spokelike fractures, termed "radial fractures," originate on the surface opposite to that on which the fracturing blow or pressure was applied. They tend to lengthen after a period of time because of internal stresses set up by the original shock. The original radial fracture will appear as a wavy line; extensions to the original fracture will run in a straight line. Temperature changes cause extensions to develop more rapidly.

As the original breaking force is continued, another force working in the opposite direction causes the glass to break in secondary fractures, termed "concentric fractures," which form a series of broken circles, or arcs, around the point of impact. These fractures are made in the opposite manner from radial fractures, the glass bending on the opposite side, then stretching and breaking on the side from which the original blow was applied. They extend from one radial fracture to another. (See Figure 8.1.)

The edges of broken pieces of glass bear a number of curved lines,

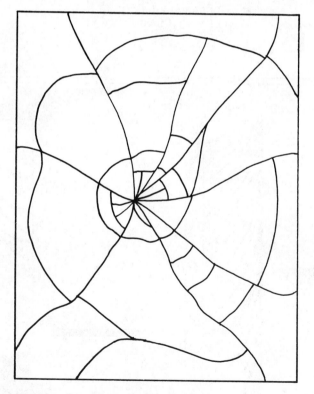

FIGURE 8.1   Radial and concentric fractures in glass

termed "stress lines." These stress lines are almost parallel to one side of the glass and perpendicular to the other. Stress lines are usually visible to the naked eye; if they are difficult to detect, the glass should be turned at various angles to the light so that the reflection will reveal them. Stress lines indicate the increase in stress set up in the glass until it breaks, and are always perpendicular (at right angles) to the side that broke first. In radial fractures, the stress lines are perpendicular to the side opposite from which the blow was struck; in concentric fractures, perpendicular to the side on which the blow was struck. From these facts the "3-R Rule" was developed: In *R*adial fractures, the stress lines are at *R*ight angles to the *R*ear surface. (See Figure 8.2.) Accordingly, if the examiner is sure

FIGURE 8.2 The 3-R rule

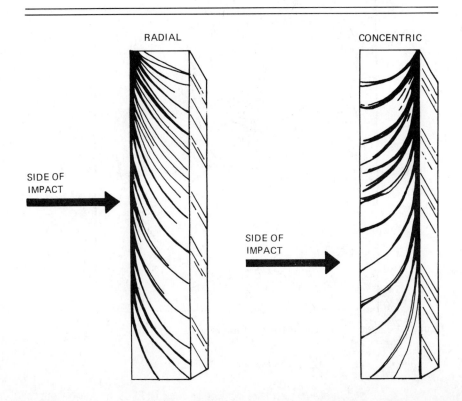

of the type of fracture he is examining, he can always tell from which direction the blow was struck.

The direction of the blow cannot usually be determined by examination of a single piece of broken glass. When it is necessary to prove the manner in which a pane of glass broke, sufficient glass fractures must be assembled to determine which are the radial edges. Determination of the outside and inside of the glass may be aided by examination of its surface, noting the amount of dirt, putty marks, lettering, and other indications that may help to replace the pane in its original form. After sufficient pieces are put back in place to enable an identification of radial and concentric fractures, the stress lines on the edges can be examined to determine the direction of the blow.

Glass fragments found at the scene of a suspected arson or other crime involving explosives or inflammables may provide the investigator with valuable information. Door or window panes broken inward may suggest a means of entry, or they may have burst from exposure to heat. Such panes broken outward may indicate the direction, force, and limits of an explosion or blast, and the point of origin. Glass fragments, especially of bottles or jars, may bear odors or chemical traces of explosive or inflammable agents, which may be either readily identifiable by the investigator or subjected to laboratory analysis for identification.

Safety Glass

Safety glass, commonly used in automobiles, consists of a transparent binding agent, such as sheet vinyl plastic, sandwiched between two sheets of ordinary glass. The binding agent prevents shattering of the ordinary glass when it is struck or receives a powerful blow. (See Figure 8.3.)

Owing to the structure of safety glass, cracking is frequently incomplete, in that neither the radial nor the concentric cracks penetrate completely from one side to the other. In such cases, the side of impact may be determined: If the concentric cracks appear on only one side, and no radial cracks are found on that side, this is obviously the side of impact; if only radial, and no concentric cracks are found on that side, this is the side away from the impact. Determination can usually be made by sliding the fingernail or a sharp-pointed instrument along the glass surface across the apparent cracks.

An additional test for side of impact is based on the property of safety glass that causes it to bend, and remain somewhat bent, instead of shattering when struck. Since the bending will result in a concave surface on the impact side and a convex surface on the opposite side, deter-

FIGURE 8.3   Safety glass (windshield)

FIGURE 8.4
Bullet hole in glass

mination may be made by placing a ruler, level, or other straightedge on the plane surface and observing the result.

Bullet Holes

Direction

The direction from which a bullet entered a piece of glass, whether window, plate, or safety, can usually be easily determined. A bullet makes a relatively clean-cut hole in the side of entrance, but as it penetrates, it pushes glass fragments ahead of it, causing a saucer-shaped depression on the exit side. (See Figures 8.4 and 8.5.) The problem is more difficult when several bullets enter safety glass close together, because the last bullets enter a glass surface that already contains a number of cracks, and as a result, small pieces are knocked out around the holes on both sides. However, broken edges on the entrance side are approximately perpendicular to the surface of the glass, whereas on the exit side, the edges are at an angle to the surface.

FIGURE 8.5   Entrance and exit of bullet

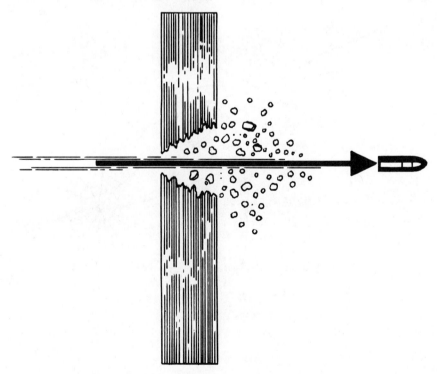

Sequence

When there are two or more bullet holes in a pane of glass, it may be possible to determine which hole was made first. A fracture in glass may travel for some distance, but if it meets another fracture that has been formed previously, it will be stopped; consequently, if fractures from one bullet hole are stopped by those of another, it may be concluded that the latter was made first. (See Figure 8.6.)

FIGURE 8.6   Sequence of bullet holes in glass

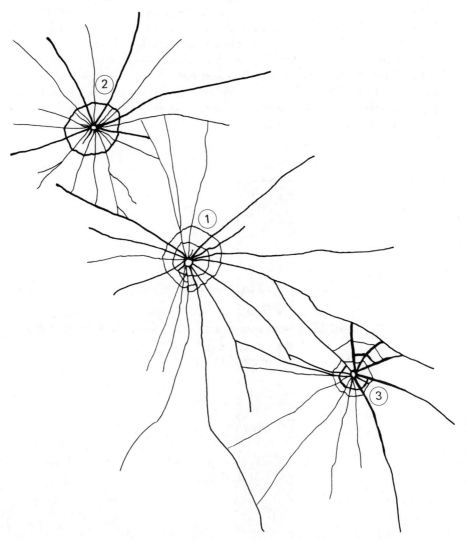

Angle

The angle at which a bullet enters a piece of glass may be determined by the amount of chipping at the exit center. If a bullet strikes glass perpendicularly to the surface, chipping around the exit hole will be fairly evenly distributed. If the bullet is fired from the right of the glass, very little chipping will be found on the right side of the exit hole, whereas there will be considerable chipping around the left side of it; also, a number of straight, short radial fractures will appear at the right of the entrance hole and one or two long radial fractures at the left of it. If the bullet is fired from the left of the glass, these indicators will be reversed. To determine the angle from which a bullet was fired, the bullet hole should be compared with test shots fired from various known angles through the same type of glass and under identical conditions, if possible —that is, with the same type of weapon and the same type of ammunition.

Type of Ammunition

The ammunition used can sometimes be determined by the size and characteristic appearance of the bullet hole. Because fragments of safety glass do not fall, bullet holes in safety glass offer more identifying information than those in window glass. When a bullet goes through a pane of glass in a sidewise fashion, it is often difficult to determine its caliber. The best method of determining the caliber and type of weapon used is to fire test shots with various weapons under conditions as nearly identical as possible to the incident at the crime scene.

Distance

A bullet fired from a long distance will have most of its velocity spent before it reaches the pane of glass, and it will break the pane in much the same way a stone will. A shot from close range with a weapon with considerable muzzle blast will produce similar results, because the blast itself breaks the glass, but it may leave powder residue and cause a crystallizing (frosting) of the glass.

Fractures Made with a Blunt Object

Glass fractures caused by the impact of a blunt instrument or object will reveal a pattern of radial and concentric fractures similar to, but not as regular as, the pattern caused by a bullet. This difference in pattern is due primarily to dispersion of the impacting force over a greater area, and more difficulty may be experienced in determining the side from which the impact came. The impact side can be determined, nevertheless, by careful examination of the stress lines on the edges of both the radial and concentric fractures.

Initially, a partial reconstruction of the object should be made, in order that radial and concentric fractures can be positively determined. On radial fracture lines, it will be seen that the portions of the stress lines on the rear side, the side opposite to the side of impact, are well developed and distinctly individual, whereas those on the front or impact side are much less so, tending to run together and lose their individuality. The "3-R Rule," as explained previously and illustrated in Figure 8.2, applies. On concentric fracture lines, the opposite condition will be found, with

FIGURE 8.7   Chipping-flaking action due to grinding and fractured edges

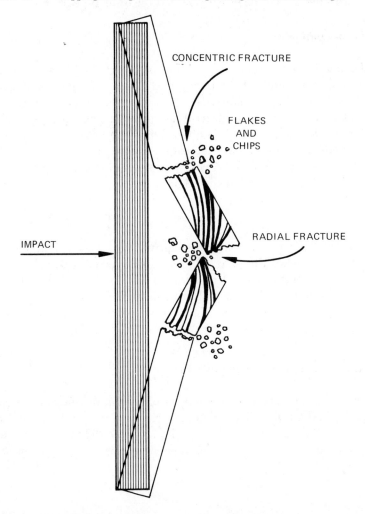

the well-developed and individually distinct portions of the fracture lines appearing on the front or impact side. This difference is ascribable to the fact that the glass bends away from the side of impact and the first (radial) fracture occurs on the rear side after the limit of stretching elasticity has been reached. This action produces the distinct stress lines on the stretching (rear) edge of the radial fractures, while the ensuing "grinding" action that takes place on the front side causes some chipping and flaking of the edge and partial obliteration of the stress lines. (See Figure 8.7.)

Fractures Due to Heat

Fractures in glass that are due to excessive exposure to heat can be distinguished from those caused by impact, since those due to heat do not show a regular pattern of radial and concentric lines, but are characteristically wave-shaped. (See Figure 8.8.) Heat fractures will also show little, if any, curve patterns (stress lines) along the edges. Expansion of the glass (stretching action) occurs first on the side exposed to the heat, and glass splinters will usually fall toward that side. Reconstruction of a glass object fractured by heat will disclose the wave-shaped fracture pattern. If the

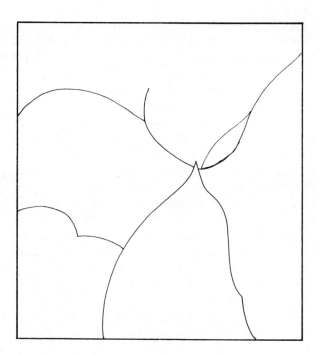

Fractures due to heat
FIGURE 8.8

stress lines are smooth, or almost so, and no point of impact or penetration is present, these factors, together with other considerations such as the circumstances under which the fragments were found and their location, may justify a conclusion that fracture was due to excessive heat.

RECONSTRUCTION OF FRACTURED GLASS

In the reconstruction of a glass object that has been fractured, one of several methods may be used, depending upon the size and shape of the object. Reconstruction should not, however, be made on any fixed or permanent basis until all examinations have been completed, nor should pieces or fragments be sent to the laboratory in a reconstructed form, because of possible damage in handling or transit.

Reconstructing a piece of fractured glass is done in much the same manner in which a jigsaw puzzle is put together. The sizes of the fragments, their shapes, and particularly the fractured edges are considered in their relation to each other, and moved around until fitting or matching edges are found. It will not always be possible to find exact and complete edges for all pieces, however, since some small fragments, chips, or flakes will either have become lost or be too small to use in the reconstruction.

For the reconstruction of a flat piece, such as a window pane, a satisfactory method is to lay the pieces on a sheet of cardboard or paper somewhat larger than the known or estimated size of the original piece. Perimeter edges can usually be identified by their straight lines and by the remains of putty or paint, or the scratches or marks left by nails or glazier's points. It is generally easier to place all these pieces first, and to reconstruct from the outside in, than to work out from the inside. Extreme care must be taken not to rub the fractured edges against one another, since this may cause further flaking or fracturing and destroy portions of the stress-line markings. A recommended method is to keep the edges a pencil point's width apart; when all available pieces are in place, the outlines may then be traced on the paper and properly annotated in accordance with the investigator's markings on the pieces, for future reference and use. If a permanent reconstruction is later desired, the pieces may be secured on a suitable base (plywood or heavy cardboard) with plastic tape or glue. (See Figure 8.9.)

Reconstruction of a curved or irregular-shaped piece, such as a bottle or jar, presents more initial difficulties, since it requires a determination—or an approximation—of both the size and the shape of the object. These difficulties are generally counterbalanced, however, by the

FIGURE 8.9
Reconstructed window pane

fact that such pieces (for example, automobile headlight lenses) frequently have patterns cast or cut into them that allow for comparisons and matching, which is not as readily possible with flat or smooth glass surfaces. In many such cases, the pattern may be matched independently of the fractured edges, although the exact matching of the edges remains the most conclusive evidence of common source.

In preparing the reconstruction of, for example, a headlight lens, the perimeter circumference size and the curvature of the spherical surface must first be determined. If sufficient pieces are available, and matching edges or patterns can be found, or if sufficient markings are available to identify the object as to make, type, size, and so forth, and permit procurement of a duplicate, the problem is simplified. A satisfactory method is to form a cast of the size and shape of the inside surface, out of linseed-oil putty or a similar material that will retain its elasticity. (If the lens can be identified and a duplicate obtained, a plaster cast may be made from the duplicate, and the fractured pieces can be mounted on it, held in place by plastic tape.) Pieces and fragments, properly marked, can then be matched by their edges and pattern markings, placed on the cast, and pressed in only sufficiently to hold them in place.

If sufficient pieces are not available to identify the object and secure a duplicate, or to assemble enough pieces to indicate the circumference and spherical surface, a piece with sufficient arcs of the circumference and sphere can be measured to a rough but sufficiently accurate degree to permit formation of the putty cast. A spherometer, if one is available, or a Geneva gauge, obtainable from an optician, can be used to determine the curvature of the spherical surface and the circumference of the lens. It must be borne in mind that these are only rough approximations, that lenses are made not only in round but also in oval and other shapes, and that the spherical surfaces are not always completely regular in contour.

If neither the spherometer nor the Geneva gauge is available, an approximation of measurements may be obtained by geometrical projection, using the arcs of the circumference and sphere of the available fragments. By tracing the arc of an available fragment on a piece of paper and using standard geometrical construction to approximate the diameter, the circumference can be projected.

PAINT SPOTS ON GLASS

Automobile glass often bears traces of the paint used on the automobile body. These traces can be of value, particularly in cases of "fleeing-the-scene" accidents, since they may indicate the color of the vehicle. Although these traces may be plainly visible to the investigator, and identifiable at least as to color, better results can be obtained in the police crime laboratory. It is essential that the investigator examine glass in such cases very carefully and that he not disturb any specks, flakes, or chips of paint, or other foreign matter on the glass. He should make specific reference to them in his request for laboratory examination and analysis.

LABORATORY EXAMINATION AND ANALYSIS

Sometimes, the investigating officer may find it advantageous to temporarily delay submission of some items, such as pieces of broken headlights found at the scene of a vehicle accident. The temporary retention of such pieces provides for possible visual matching of remaining pieces found on a suspected vehicle, to be later confirmed, if possible, by laboratory examination and analysis.

Conversely, early submission for laboratory analysis may aid in the investigation, as in the case of an accident where no suspect vehicle is immediately located. The laboratory may, if enough glass is available, identify the vehicle as to make, model, and year of manufacture, or eliminate from consideration glass from a suspect vehicle. Since glass is not destroyed or appreciably altered in laboratory examination (except in spectrographic examination of small fragments), the evidence pieces are available for later comparison with suspect pieces.

The information in the following paragraphs with reference to laboratory methods of analysis is not intended to make the investigator an expert in this field, but it will provide him with a general working knowledge of various methods of analysis that qualified laboratory technicians can use to assist him in his investigation of glass and glass fragments.

Glass Fragments

Various glass fragments, such as broken headlight lenses, broken bottles, and similar glass materials, may furnish important investigative leads when examined by properly trained technicians to determine their composition and possible identity, or nonidentity, with other fragments. A scientific examination of several particles of glass may disclose identical physical and chemical characteristics, indicating that all particles came from the same piece of glass. It can be determined whether minute particles having the physical appearance of glass are actually made of glass.

Crime laboratory analysis of glass or fractured glass located at the scene of an incident may indicate:

1. The type of glass
2. The manufacturer of the glass
3. That the glass did or did not come from a certain place, thing, or item
4. The direction of the blow
5. The direction and angle of impact of the bullet, instrument, or object
6. The sequence of holes

Fluorescence of Glass

Examining glass fragments under ultraviolet light may not produce results as positive as those produced by other examinations. It may, however, be useful in determining whether two pieces of glass could or could not have come from the same object. The fluorescence examination is based on the fact that the mineral constituents will give a distinctive type and degree of fluorescence to the glass originating from one molten batch. Although the results of ultraviolet comparisons of two or more pieces may not be positive in establishing similarity, they may be positive as to dissimilarity, and thus useful in the elimination process of an investigation.

Since fluorescence examination requires that the glass be absolutely clean, to preclude fluorescent reaction from contaminants, it must be washed in acetone or a similar solvent, and also in water. Therefore, caution should be exercised that this examination is not performed until after other examinations for fingerprints, surface debris, and so forth, have been completed.

Spectrographic Analysis

The constituents of glass, especially the minor or trace constituents such as contaminants and minor elements added for specific purposes, lend themselves readily to spectrographic analysis, which identifies and measures the quantities of these constituents. The presence or absence, and the quantities, of these minor constituents are of more importance in the spectrographic analysis than are the basic constituents, which are ordinarily present in such comparatively great quantities that differences in the spectral lines of various fragments are often difficult to detect. The spectrograph, for this reason, is of greatest value in demonstrating major differences between two samples, indicating their nonidentity, rather than in indicating acceptably proven identity.

Spectrographic analysis should not be routinely requested in all cases, since it is not always necessary or useful to the investigation, and since small fragments are destroyed by the analysis. Other examinations should be completed first, and spectrographic analysis requested only if it is considered necessary.

Refractive Index

Refraction refers to the change of direction of a ray of light in passing through a medium whose density is not uniform, as through a piece of glass that has been ground to specific requirements. The refractive index is an essential characteristic of such items as optical glass and of minor consideration in other types, since no attempt is ordinarily made in the manufacturing process of the latter to control it to specific measurements.

The refractive index can be measured either microscopically or with a refractometer; however, the latter method requires that the two sides of the sample be parallel. Another method, known as "flotation," in which a chemical solution is used in conjunction with the refractometer, can be used when the specimen does not have parallel sides.

Density

This analysis is based on the relative densities of two or more pieces of glass. Comparison of the densities of a known and an unknown piece of glass may aid in determining if they are similar or dissimilar.

FORWARDING TO THE LABORATORY

It should be the responsibility of the investigator to prepare and ship (or carry) the evidence to the laboratory. The evidence must be packed in such a manner as to prevent breakage, friction, shifting, or contact with other items that would result in destruction, loss, or contamination of the evidence being forwarded.

All available pieces and fragments pertaining to one incident should be submitted at the same time. To facilitate the work at the laboratory, each piece of evidence should be identified clearly on individually wrapped items.

9

Questioned Documents and Related Offenses

INTRODUCTION

This chapter provides the investigator with information that will enable him to efficiently investigate incidents involving questioned documents. It covers the important aspects of his examination of a document, and the procedures to follow to obtain exemplars from which the laboratory document examiner will be able to reach a sound conclusion. It will not cover any laboratory procedures except insofar as it may be necessary to clarify certain points.

DEFINITIONS

Document: Any material that bears handwriting, handprinting, print, typewriting, or any kind of drawing or mark. "Handwriting" includes any and all systems by which words, letters, symbols, or characters are produced on any type of surface, such as walls, blackboards, cardboard, or paper of any kind or description, as long as the graphic representations are made by hand.

Questioned document: A document whose genuineness is questioned, normally because of origin, authenticity, age, or the circumstances under which it was written.

Holographic document: A document that is wholly in the handwriting of one person.

Standard: A document recognized as proven, genuine, or acknowledged that has been obtained from official records, personal letters, etc., and is known to be the product of a particular person or machine.

Exemplar: A document requested by the investigator that duplicates the text of a questioned document and is known to be the product of a particular person or machine.

INVESTIGATION

Questioned-document cases arise in the course of everyday business. All records and forms are vulnerable to tampering, and may become evidence in a case under investigation. Suicide notes may be examined to determine if they are what they purport to be; pay slips, legal documents, anonymous letters, and threatening and obscene letters are some of the other documents with which the investigator may be concerned.

Questioned documents fall into two general classifications: genuine and fraudulent. It is often of greater importance to prove a document genuine than it is to prove it forged. A genuine document is often disputed because it imposes a liability on the maker. For example, a person who writes a genuine check may later deny that it is genuine in an attempt to avoid financial responsibility.

Offenses that involve questioned documents, such as forgery and the alteration of documents—including the making and passing of bad checks and bogus bills of lading—are often recurring offenses. The offender may repeat his crime many times, using the same method of operation. The investigator should therefore give special attention to factors that may link the offense under investigation to other, similar incidents.

The investigator must recognize the importance of the laboratory document examiner in processing evidence in fraudulent-document cases. A good investigator should be familiar with the services the laboratory can provide him.

The techniques used in laboratory processing are applicable to writings in any language, using any alphabet; the document examiner does not necessarily have to be acquainted with an alphabet before he can reach a firm scientific finding. Document examination is based on the improbability of any two writings being exactly alike in all characteristics, such as style, speed, slant, and spacing. (See Figure 9.1.) Since writing involves a mental process, regardless of the skill and habitual performance, as well as muscular coordination, the basic principles of handwriting identification hold true even if all the people of the earth learned to write from one standard copybook.

In addition to the physical characteristics mentioned above, there are other clues to help identify a particular writer. The terminology used in a letter, or referral to an incident that may not be generally known, may provide an insight into the identification of the writer. The writer can often be identified with a particular group by the use of highly technical terms. Punctuation, spelling, grammar, syntax, and style are all valid clues that may lead to the writer's identity.

FIGURE 9.1 Microscopic examination of submitted document

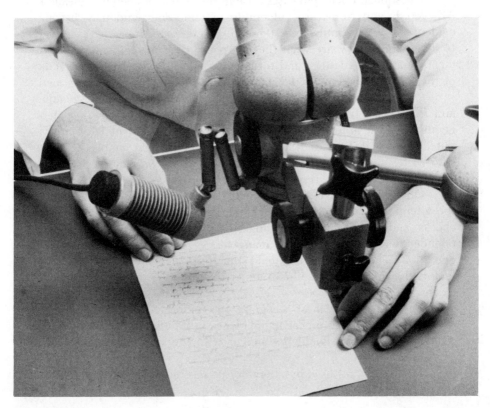

OFFENSES INVOLVING QUESTIONED DOCUMENTS

Because questioned-document cases arise in many different contexts, the investigator must be able to distinguish between the various offenses that may be involved, in order to provide the evidence necessary to prove each offense. The offenses discussed below are those that occur most often in cases involving questioned documents.

Forgery

Forgery is an attempt by a person to defraud another by falsely making or altering a signature or any part of a document that would, if genuine, appear to impose a legal liability on another or to change the victim's legal rights or liabilities to the offender's illegal advantage. Forgery is also committed by offering, issuing, or transferring such a writing in an attempt to defraud another, provided that the offender knows the document to be falsely made or altered. The term *false*, as used in defining forgery, does not refer to the truth of facts stated in the document; it refers only to the documents not being genuine. For example, a check that is drawn on an account containing insufficient funds but signed by the maker in his own signature, even if made with the intent to defraud the bank, is not forgery; it is the offering of a "bad check." If the maker signs the name of another real person, however, the signature is falsely made. Similarly, a document that has been altered so as to change the amount of indebtedness it represents is falsely made, because it purports to represent a debt different from the one originally intended.

To prove a case of forgery, the investigator must gather sufficient evidence to show:

1. That a certain document appears to have been made by a person other than the true maker, or has been altered to represent a different obligation from that intended by the maker
2. That the making or alteration of the document was not authorized by the apparent obligor
3. That the document was of a nature that would, if genuine, appear to change a legal right or obligation to the prejudice of the apparent obligor
4. That it was the accused who falsely made or altered the document, or offered, issued, or transferred it
5. That the accused knew that the instrument was falsely made or altered at the time he offered, issued, or transferred it
6. That the intent of the accused was to defraud

Checks on Insufficient Funds

"Bad check" offenses include making, offering, or transferring any check, draft, or order for the payment of money upon a bank or other depository in order to procure anything of value by fraud, or to pay any past-due obligation. This offense will often apply when forgery will not. For example, a check drawn in the name of a nonexistent person is not forgery; it is a "bad check" offense. To prove a "bad check," the evidence must show:

1. That the accused made, offered, or transferred a check, draft, or order payable to a named person or organization
2. That he did it for the purpose of procuring something of value, or for the purpose or purported purpose of making payment for a past-due obligation
3. That the act was committed with intent to defraud or deceive
4. That the accused knew at the time he did the act that the instrument would not be paid on presentment

Extortion

Extortion is the communication of threats to another person with the intention of obtaining anything of value, or any acquittance, advantage, or immunity. (See Figure 9.2.) An *acquittance* is a release or discharge from an obligation. Proof must show:

1. That the accused communicated certain threats to another
2. That the threat was received by the intended victim
3. That the accused intended to obtain something of value, or any acquittance, advantage, or immunity by unlawful means

INVESTIGATIVE CONSIDERATIONS

Investigator's Examination

Preliminary Examination

In order for the laboratory document examiner to be able to reach conclusions of a demonstrable nature, he must be furnished adequate and

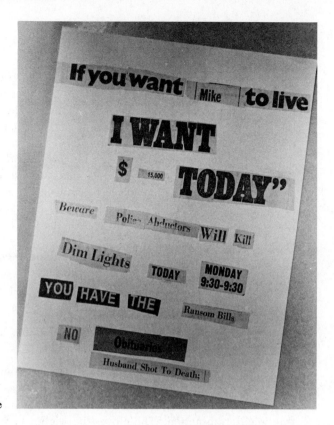

FIGURE 9.2
Extortion—a ransom note

proper exemplars with which to compare the questioned document. And in order for the investigator to be able to provide as much assistance in this aspect as possible, he must know what is required. The only way he can determine this is by a careful examination of the questioned document. Whenever a case involves the origin or validity of a document, the investigator should attempt to contact the document section of the laboratory by telephone and, if possible, receive an answer prior to forwarding the case to the laboratory.

Identification

Initial. The first act of the investigator should be to appropriately mark the document so that he can identify it at any later date. The document should be examined to determine a suitable place where the investigator can enter his initials, the date, the time, and the case number, if it is known. This location should be chosen with care, and the identification date should be as inconspicuous as possible. The investigator's identification should not in any way interfere with any writings that are on the document. A corner on the back of a document is the place most commonly used.

For Transmittal to the Laboratory. When transmitting questioned-document evidence to the laboratory, in addition to the normal marking for identification above, the investigator marks the evidence by means of the "*Q*" and "*K*" system. This consists of, first, marking each questioned document with an uppercase *Q* and each known document with an uppercase *K*. This immediately identifies *questioned* as opposed to *known* documents and greatly facilitates the laboratory examiner's work.

Then, each individual piece of evidence is numbered consecutively under the questioned (Q) and known (K) categories. For example, the first questioned document would be Q1; the first known document would be K1.

Finally, copies are identified by marking them with a lowercase *c*. Copies include photographic, heat reproduction, carbon, and others. For example, if the fifth questioned document in a case is a carbon copy, it would be marked "Q5c."

Receipt of Acceptance

The documents involved in questioned-document cases often represent valuable transactions. A document can be used by the victim as evidence in a civil suit to recover losses he has suffered because of the fraudulent transaction. Therefore, the victim will need assurance that the document will be returned after the investigation is completed. The investigator should give a property receipt to the person who gives him the document, just as he would if he took other types of property. The receipt should describe the document in sufficient detail to permit future identification. However, it should not contain positive statements as to the value of the document; it should be limited to a physical description. Similar receipts should be given for any related property collected, such as pens, pencils, or paper.

Notations for Future Reference

The investigator should take notes concerning his collection of the document. They will aid him later in refreshing his memory if he is called on to testify. He should note the place, time, and date that he collected the document, the name of the person from whom he received it, and how the document was marked. He should also include information about the history and contents of the document. Later, he may want to add notations about his handling and disposition of it. All this information may be important later in establishing the law enforcement agency's case.

Care and Preservation

A questioned document should never be folded, crumpled, or carried unprotected in a pocket. It should be placed between transparent covers, to permit handling for subsequent examinations without direct manipulation, thus preventing any marring by contamination or destruction by abrasive action. If the document is torn, no attempt at restoration should be made, other than placing the pieces in the protective covering in their most obvious and logical positions. If the document is to be examined for fingerprints, it should be handled with tweezers to prevent the addition of the fingerprints of the investigator.

The document should not be subject to strong light for prolonged periods. Ultraviolet light can, however, be used to discover information not visible under normal light.

It is often desirable to make copies of the questioned document for use during the investigation. The original may then be placed in the evidence room until required by the laboratory for examination. Reproduction procedures that require removing the document from the protective covers should be avoided; Xeroxing and photographing are acceptable methods for making copies.

Interviewing the Victim

The number of victims in any questioned-document case may vary, depending on the type of document involved. The investigator should question everyone affected by the document. For example, in a case involving a forged check, the investigator will need to question the person who cashed the check, the person whose signature has allegedly been forged, and personnel at the bank on which the check was drawn.

The investigator should attempt to find out how and when the document was prepared or used. He should get a description of the suspect, of his appearance, actions, and conversation, and of any credentials he may have used. The number of suspects and the number of people present when the document was offered or found are also important. He should find out how it was discovered that the document was false or why it is suspected of being false.

If the document was prepared or signed in the presence of a witness, the investigator should ask the witness about the method of preparation. Was it written with the right or left hand? Was it written quickly or slowly? Was it written on top of other papers, or on a hard surface? Did the writer seem nervous or intoxicated at the time?

If the signature is that of a real person, he should be interviewed to verify that he denies writing or signing the document. He should be permitted to examine the document, but the investigator must ensure that no one is given an opportunity to damage or destroy the evidence. Victims should always be treated as suspects until the document expert determines that the victim did not make the document. Therefore, it is necessary to warn the victim of his rights before taking his statement.

If the questioned document is written on a standard form, the investigator may find it useful to talk to those who normally use the forms, to examine the area where they are kept, and to find out who has access to the area.

Any bank, business, or other organization that will be affected by the questioned document should also be contacted. Information about its past dealings with the person whose signature has allegedly been forged may offer helpful clues, and similar incidents, using the same forms or methods of operation, may be discovered.

All victims and witnesses should be encouraged to name possible suspects. The investigator should ascertain the reasons for these suspicions. After completing a list of all the victims and suspects, the investigator may find it helpful to look into their financial status and business practices to determine possible motives for the offense. Emotional stability of the victims and suspects may also be useful indicators. The investigator may then attempt to reduce the number of subjects from whom voluntary exemplars and collect standards must be obtained.

Exemplars

The investigator must furnish the laboratory technician exemplars to compare with the questioned document, so the technician may determine the identity of the document's author. The degree of accuracy of the laboratory technician's conclusions depends on having exemplars that duplicate the questioned document as closely as possible, and on having a sufficient number of exemplars for a realistic comparison.

Paper

During the investigator's preliminary examination, he should have determined the type of paper comprising the questioned document. If it is on a standard form, he must obtain samples of that form or have facsimiles reproduced. If the document is a check, samples of the same type of check, and of the same size and quality as the questioned check, must be obtained. Samples of other types of papers must be obtained to approximate that of the questioned document as closely as possible. The paper may be bond or onionskin, colored or white, ruled or unruled, in tablet form, wrapping paper, or a piece of cardboard. The same type of material must be obtained as that used for the questioned document.

Writing Instrument

The investigator's preliminary examination should have revealed to him the type of instrument used to write the questioned document. A similar instrument should be obtained. If the instrument was determined to be a ballpoint pen, then a ballpoint pen with a similar size ball should be used to write the exemplars. If the writing was produced with a pencil, then a pencil should be used that has the same degree of hardness and is sharpened to approximately the same point as the one that produced the questioned writing. Other writing instruments, such as crayons, nib pens, brushes, or those designed for special purposes, should be obtained as appropriate. If the suspect owns a writing instrument with characteristics similar to the questioned instrument, at least one exemplar should be obtained with it.

Confinement of Writing

When questioned writing appears on a form, the writing is usually confined to a limited area by means of blocks, length of lines and their emplacement, or the size and design of the form. In all cases, exemplar writings must be similarly confined. For example, if the questioned writing appears on a form, duplicates of which cannot be obtained, then the investigator must provide paper of the approximate quality, cut to the same size as the questioned form, and with lines or similar confining devices drawn on the paper to simulate those of the questioned form.

Examination of Standards

The investigator should obtain standards to familiarize himself with the suspect's normal writing before he attempts to obtain exemplars from him. This will enable him to determine, at least superficially, whether the suspect is attempting to disguise his handwriting. The general pictorial quality of the writings should appear similar if no attempt to disguise the writing is undertaken. The difference between a normal slant and a backhand, or the difference between letters written in a round style as opposed to angular writings, is generally quite obvious. Disguised writing affects the pictorial appearance only.

Warning

The investigator must advise the suspect that exemplars furnished by him must be voluntary, that they can be used against him, and that he has a right to refuse to execute them. He must also be told the nature of the offense involved. Prior to taking specimen writings, a handwritten statement should be obtained from the suspect indicating that his handwriting exemplars are in fact voluntarily executed. After the suspect has signed the statement, the investigator should sign or initial it and date it for later identification.

Writing the Exemplars

Exemplar writings should be obtained from the suspect by dictating to him what he is to write. The suspect should not be permitted to copy precedingly written phrases; that is, each line must be written without seeing the preceding line. No indications of spelling or punctuation should be given. If the questioned document is lengthy and contains no unusual spellings or grammatical constructions, the material that the investigator wishes the suspect to write may be typewritten for the suspect to copy from. The suspect should never be shown the questioned document prior to obtaining the exemplars.

In cases where the questioned writing is a signature, about 25 exemplar signatures are required. Where one or two sentences, or about 20 disconnected words, comprise the questioned writing, about 10 to 15 exemplars are required. If the questioned document consists of lengthy writing of two or three pages, one exemplar should be obtained of the entire text, then about three to five exemplar writings should be obtained consisting of the first two or three paragraphs and last two or three paragraphs of the questioned writing. If any words in combination with each other, or unusual wordings, are found elsewhere in the questioned document, they should be included in the exemplar writings.

When the suspect has completed an individual exemplar, it should be immediately removed from his sight. This will preclude an attempt to copy the previously made exemplar.

Handwriting exemplars should be obtained from the victim when the questioned document is an anonymous letter.

Exemplar Appearance

The pictorial quality of a writing may enable the investigator to determine that the exemplars are in fact truly representative of the writing of the suspect. This may be determined in several ways. Smooth, unbroken strokes and rounded forms of letters are usually indicative of speed in writing, and rapid writing will generally reduce the size of the letters. If the exemplar writing appears to be written slower or faster than the questioned writing, then the investigator should increase the speed of his dictation, or request the suspect to write faster or slower, as the case may be.

A suspect's visual memory, powers of graphic expression, and manual dexterity are often indicated by the general appearance of his writing. The investigator, by comparing the standards, the exemplars, and the questioned writing, may be able to determine continuity in the degree of skill of which the writer is capable. Skillful writing is fluent and mature, differing to a large degree from the copybook style of the unskillful. Skill and line quality go together. The lines of skillful writing will be freely written and uninterrupted, the curves will be well defined, and the pauses and pen lifts will occur in natural ways and in appropriate places.

The investigator should consider obtaining exemplars written with the left hand if the suspect is writing with his right hand, or vice versa. If the writer is unable to produce legible writing with other than his normal hand, one exemplar is sufficient. Exemplars may be obtained with the suspect writing standing up, or holding the paper against a wall, if it is deemed desirable.

Precautions

Exemplar writings should be on only one side of the paper provided. In a case involving a check where the writing on both the face and the back are questioned, sufficient blank checks should be obtained so that exemplars of the writing appearing on the back of the questioned check will be written separately, on different checks, from the exemplars of the writing on the front.

Exemplar writings of questioned documents containing obscene words or phrases may be produced without the objectionable words or phrases if the document is of sufficient length so that the elimination of such words will not render the exemplars worthless. Exemplars of short obscene documents will usually have to be produced in their entirety. The investigator is faced with a similar problem when the questioned document contains security or politically sensitive information. If the exemplar can be produced by omitting the sensitive portions, it should be so produced.

Handwritten Standards

Standards are those specimens of handwriting produced in the normal course of events. These may be found in official files and consist of documents and forms required by law to be in the handwriting, or bear the signature, of the person filling them out in compliance with such law. Letters previously written by the suspect may also be used. One of the legal requirements that must be met before these documents will be accepted in court as evidence is that they be proved official and genuine. Three methods of proof are:

1. The personal acknowledgment of the people providing the standards
2. Testimony from witnesses who can state from their own knowledge that certain people provided the standards
3. By comparison with exemplars known to have been written by a certain person

It is recommended that all standards obtained by the investigator be discussed with the district attorney as to their admissibility and proof of genuineness. When the genuineness of a signature is questioned, standards of a known genuine signature should be obtained, especially those that may have been in the possession of, or accessible to, a suspect. This may permit the laboratory document examiner to determine whether the genuine signature has been traced and, if so, to determine which actual signature was traced. In all cases when exemplars are obtained, whether it be from the subject or victim, standards should be submitted together with the collected exemplars. This permits the examiner to determine the normal handwriting habits of the writer, and also indicates attempts to disguise the writing in the exemplars.

Examination of Typewritten Documents

Often, a large number of typewriters must be examined to determine which one produced the questioned document. The investigator can eliminate many of the possible typewriters without the necessity of submitting exemplars from them to the laboratory. He can accomplish this by several relatively easy field tests. The upper- and lowercase letters M and W should be examined first, as they are usually the most distinctive as to style, and differences may be more easily recognized by the investigator. (See Figure 9.3) The bottom of the staffs of the lower case m may or may not have serifs (short lines at the ends of strokes), or the two outside staffs may have serifs and the center staff none. The center V-like formation of the capital M may descend to the base line or terminate varying distances above it, and it may or may not have a serif. The inverted V of the center formation of the W may or may not extend to the top of the line formed by the outer portions of the letter and may or may not have a serif at the top.

FIGURE 9.3
Letter differences

The numerals are also distinctive in design on various typewriters and should be examined by the investigator. He should have no difficulty when different sizes of type are involved. Unless the letters and numerals examined by the investigator are distinguishable with ease, exemplars should be submitted to the laboratory. (See Figure 9.4.)

Typewritten Exemplars

Several exemplars must be obtained from each machine if a successful identification is to be made. Unless, through excessive abuse, a typewriter produces typing of extreme individuality, identifying writing as having been produced by one machine is time-consuming and requires the use of precision instruments. (See Figure 9.5.)

Also, each typist develops certain individual characteristics of typing

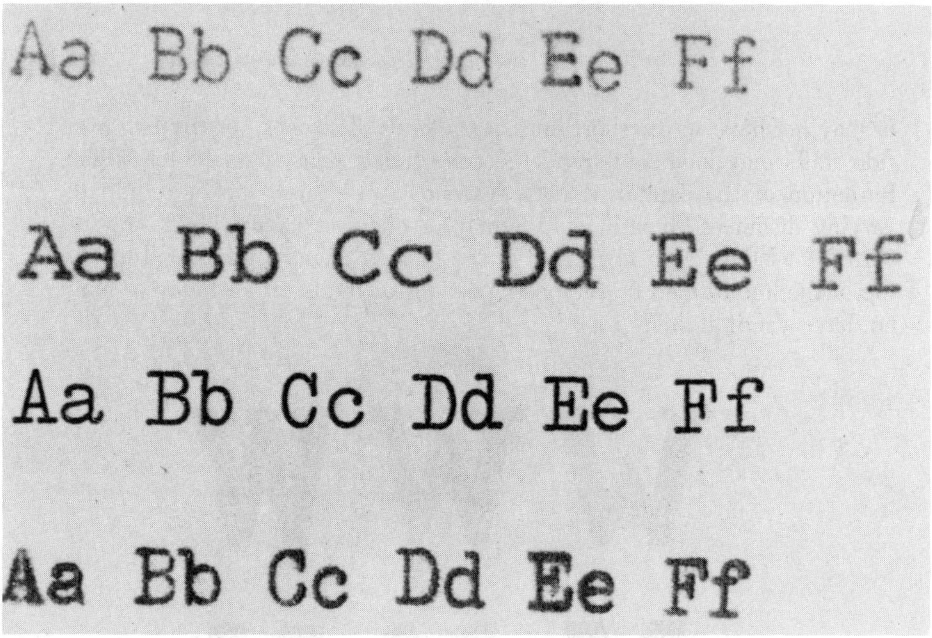

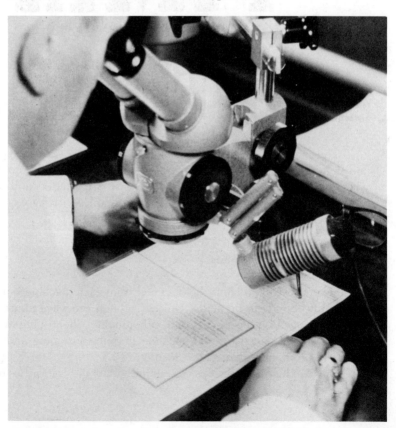

FIGURE 9.4 Typewriting characteristics

FIGURE 9.5 Laboratory examination of typewritten

that can, in many instances, be identified. That is, work produced by different typists can often be differentiated by their individually developed techniques, talents, coordination, skill, and aptitude. These individual characteristics may result in such things as heavier impressions of some letters as opposed to others (this will not be true of electrically operated typewriters), individuality in spacing methods, and two letters in combination consistently indicating a rhythm differing from the rhythm of other letter combinations. The investigator should attempt to obtain an exemplar written by the person who normally operates the typewriter in question.

All exemplars should have, as a heading, the name of the person producing the exemplar, the make of typewriter and its serial number, the place where the exemplar was produced, and the date. The signature of the person producing the exemplar and that of the witnessing investigator should be included for identification purposes.

The investigator should make sure that the exemplar follows the style of the questioned document. If the questioned typewriting is double-spaced, the exemplar should be, too. Indentions should be the same. The length and number of words in each line should be the same as in the questioned writing. The word or portion of a word at the end of a line on the questioned document should be duplicated in the exemplar. The exemplar should be as close a duplicate of the questioned document as it is physically possible to produce.

A second exemplar should be typed by someone other than the person typing the first one. In addition, the second exemplar should have the complete keyboard reproduced, first in normal order, and then with a space after each letter, numeral, and symbol. If the questioned document is a carbon copy, a carbon-copy exemplar should be included.

A third exemplar should be made by the carbon-stencil method. A new piece of carbon paper should be used. It is placed in contact with the paper as though the paper were to be a carbon copy; that is, with the carbon side against the paper to be used to obtain the carbon stencil. The ribbon of the typewriter should be removed or the machine set on stencil. The type should strike the carbon paper directly. The spacing, line length, and word composition should be the same as in the other exemplars. The carbon paper need not be submitted with the exemplar. The carbon-stencil exemplar may be omitted if the typewriter normally uses a carbon-paper ribbon.

If the questioned document consists of about half a page, it should be reproduced in its entirety. If the document is lengthy, the first 20 to 30 lines should be reproduced. The remainder of the questioned document should then be examined and any words, numerals, or symbols not appearing in the first 20 or 30 lines should be added to the exemplar. The words preceding and following the material to be added should be included and typed as they appear in the questioned document.

The investigator should compare the exemplars and the questioned document to ascertain whether the ribbon producing the exemplars appears to have about the same degree of wear and is of the same style and character as the one that produced the questioned document. Information should be obtained as to when the ribbon on the machine was last changed, and the date and nature of the latest repair work performed on the typewriter.

Typewritten Standards

When it can be determined that the questioned document was typed on or about a certain date, an effort should be made to obtain standards of the work typed on that particular machine during that period. A comparison of the typing produced by a particular machine during a particular period of time with the typing produced by it at a later date will reveal any additional defects, flaws, or changes in its typewriting characteristics.

Transmittal to Laboratory

Transmitting document evidence to the laboratory does not require elaborate preparation, but there are some precautions to be taken. The questioned document must be protected from damage. This is best accomplished by placing sufficiently heavy wrapping material around it to prevent it from being bent, torn, or folded in transit. The questioned document must be identified, so that it will not be confused with any of the other documents submitted. The laboratory request should be written so that the laboratory examiner is not given any clue as to which of several subjects the investigator suspects to be the author. Even though all laboratory examinations are made without bias, the wording of the request may influence a court to believe otherwise if this is made an issue.

MISCELLANEOUS DOCUMENT EXAMINATION

Indented Writing

One of the relatively infrequent but important examinations the laboratory is called upon to perform is that of the decipherment of indented writing.

When the writing on a document has been obliterated, accidentally or otherwise, or when the only evidence available is second sheets (sheets of paper underlying the one on which the original writing was placed), the pressure of the writing instrument will have left a trace of the writing on the document or the second sheets. This is called *indented writing*. If the pressure is great enough, the indented writing might even be obtained from the sheets underlying the original by several sheets. The laboratory can frequently decipher indented writing by utilizing chemical processing, special lighting, and photographic techniques.

Charred Documents

The investigator should understand the difference between a charred document and ashes. A *charred document* is one that has become blackened and brittle, having been burnt in an absence or an excess of oxygen. When documents are inside a container that has been involved in a fire, the heat will often cause them to catch fire, but since there is very little oxygen present, the paper will char instead of being consumed and turned into ash. No restoration of ashes is possible, but charred documents are frequently treated so that the writing on them can be made visible, usually by infrared photography.

Charred documents are extremely fragile and, if not handled carefully, may well be destroyed beyond redemption. If it is at all feasible, the charred documents should be sent to the laboratory by courier, to preclude unnecessary handling and prevent destruction. In some cases, the laboratory technician should be requested to come to the location of the document. If neither of these two preferred methods is practical, careful packaging must be resorted to, to preclude destruction.

Mutilated Documents

Documents that have been torn, accidentally or purposely, or mutilated by being subjected to the actions of a washing machine, may be restored with care and the expenditure of time. The investigator should not attempt to assemble mutilated documents by taping the pieces together or by gluing them to some other piece of paper or other backing. Success is doubly assured if the documents are sent to the laboratory in the condition in which they are recovered.

Inks

Samples of ink that appear to the unaided eye to be identical may be determined by laboratory examination to be entirely different in composition. One technique of laboratory examination is chromatography, a chemical process that separates the compounds into separate colored layers. Very little ink is required for this process; a single capital letter may furnish a sufficient quantity.

However, the removal of this ink from a document may affect the admissibility of the document in court. Therefore, the proposal to use this technique should be discussed beforehand with the district attorney. If he recommends against it, it should not be used. If he decides that the removal will not affect the document's admissibility, the laboratory technician should be so advised.

Paper

Paper samples can be analyzed in the laboratory to determine similarity or differences in composition that may lead to a conclusion concerning a common origin with other samples. Adhesive used to hold sheets of paper in tablet form can be analyzed to determine whether two or more sheets of paper could originally have been a part of one tablet. The investigator may examine papers under ultraviolet light to detect similarities or differences in fluorescence or reflectance that may provide him with an investigative lead pending the laboratory analysis.

10

Drugs

INTRODUCTION

This chapter provides criminal investigators with information on the characteristics of the most widely used drugs, and methods and techniques used in the investigation of offenses involving the possession, sale, or use of contraband narcotics and other drugs.
Figure 10.1 illustrates many of these.
The Comprehensive Drug Abuse Prevention and Control Act of 1970, which was signed into law by the president on October 27, 1970, provides a more useful and meaningful basis from which law enforcement officers, researchers, and legal and medical communities can work.

DEFINITIONS

Narcotics: Any opiates, synthetic opiates, or cocaine. Examples are opium, heroin, morphine, codeine, paregoric, Dilaudid, pethidine, and methadone.

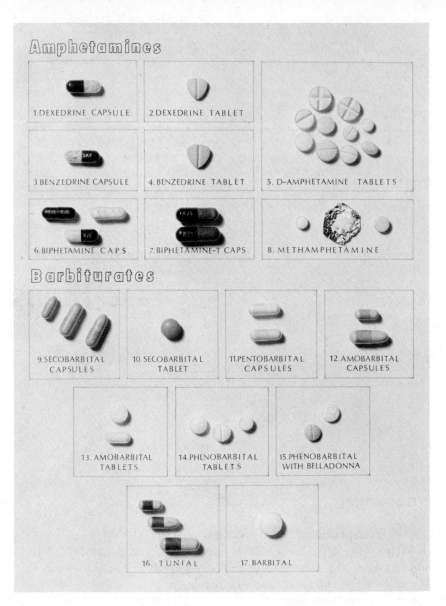

FIGURE 10.1   Drug identification chart

Marihuana: The intoxicating products of the Indian hemp plant, *cannabis sativa,* or any synthesis thereof, including hashish. Tetrahydrocannabinol (THC), the active ingredient of marihuana, is a strong hallucinogen, with sedative properties.

Dangerous drugs (an administrative label): Those nonnarcotic substances that the attorney general or his designee, after investigation, has found to have, and by regulation designates as having, a potential for abuse because of their depressant or stimulant effect on the central nervous system or their hallucinogenic effect. (Pl. 91-513)

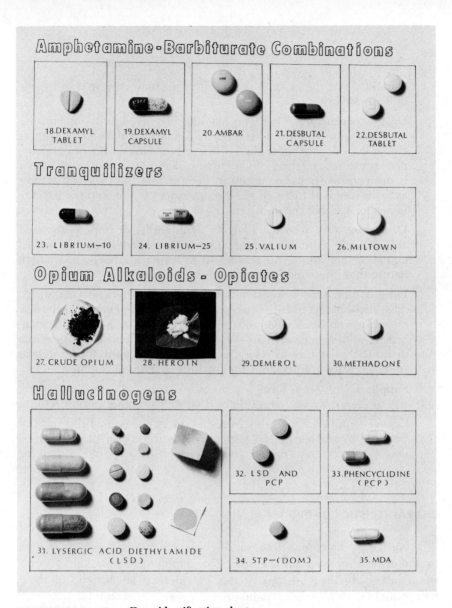

FIGURE 10.1 (cont.) Drug identification chart

Drug abuse: The illegal, wrongful, or improper use of any narcotic substance, marihuana, or other dangerous drug, or the illegal or wrongful possession, sale, transfer, delivery, or manufacture of the same. When such drugs have been prescribed by competent medical personnel for medical purposes, their proper use by the patient is not drug abuse.

Drug abuser: One who has illegally, wrongfully, or improperly used any narcotic substance, marihuana, or dangerous drug, or who has illegally or wrongfully possessed, sold, transferred, delivered, or manufactured the same. A drug abuser is usually referred to by one of the next three terms.

Drug experimenter: One who has illegally, wrongfully, or improperly used any narcotic substance, marihuana, or dangerous drug as defined above, not more than a few times, for reasons of curiosity, peer pressure, or other, similar reason. The exact number of usages is not necessarily as important in determining the category of user as is the intent of the user, the circumstances of use, and the psychological makeup of the user.

Drug user: One who has illegally, wrongfully, or improperly used any narcotic substance, marihuana, or dangerous drug as defined above, several times, for reasons of a deeper and more continuing nature than those motivating the drug experimenter.

Drug addict: One who exhibits a behavioral pattern of compulsive drug use, characterized by overwhelming involvement with the use of a drug and the securing of its supply. As the term is used herein, a drug addict may or may not be physically dependent upon the drug. Rather, the term refers in a quantitative sense to the degree to which drug use pervades the total life activity of the user.

Supplier: One who furnishes illegally, wrongfully, or improperly any of the proscribed drugs defined above to another person.

Casual supplier: One who furnishes illegally, wrongfully, or improperly to another person a small amount of any of the proscribed drugs defined above for the convenience of the user rather than for gain.

CHARACTERISTICS OF DRUGS

Narcotics, dangerous drugs, and marihuana will be categorized here by the effect they have on the user. Four groupings result: depressants, stimulants, hallucinogens, and volatile chemicals. The purpose of these four groupings is to enable criminal investigators to become familiar with the appearance, composition, and physical effect of these substances. Commonly found user terms and terminology have been added as further assistance in becoming familiar with the problem.

The Depressants

General depressants act upon the central nervous system and slow it down. The term *depressant* is often associated with the feeling of being depressed, a dejected state of mind, or anxiety; for this reason, the use of the word sometimes causes confusion. The drug abuser would argue against using the term *depressant*. The "high" that drug users constantly strive to attain is a state of well-being, release, and comfort—an escape mechanism from the reality with which they cannot or will not cope. This imagined or real state of mind is called *euphoria*. Seeking the new or better high is what motivates many people to move from drug to drug.

Opium

Opium is the milky-white juice that is obtained from the pod of the poppy plant, *papaver somniferum*. The substance darkens to a blackish-brown color and thickens upon exposure to air.

Raw opium has a distinctive and pungent odor. It is most commonly used by smoking in long-stemmed pipes, but there have been reported instances in Vietnam and Korea of marihuana cigarettes being dipped in opium solution and smoked. The user presents a sleepy and relaxed appearance while under the influence of opium.

Opium creates both mental and physical dependence in the user. It is the source of morphine, heroin, codeine, paregoric, and other derivatives.

Morphine

Morphine is obtained from raw opium base through a chemical process in which ten pounds of raw opium yields one pound of morphine base.

Morphine usually appears as an odorless white crystalline powder on the illicit market but also appears in tablet, capsule, and liquid form. It provides medical standards by which other narcotics are evaluated.

Morphine is usually administered by injection with a hypodermic syringe (or a combination of hypodermic needle and medicine dropper), and creates both mental and physical dependence in the user. A euphoric state is obtained with morphine, and the user presents a sleepy or relaxed appearance and may exhibit constricted pupils of the eyes.

Heroin (Diacetylmorphine)

Heroin, like morphine, usually appears as a crystalline white powder, odorless, and normally sold in glassine paper packets or in capsules. Heroin that is darker in color contains more impurities.

Because heroin is about four to five times stronger than morphine, it is normally the drug addict's choice among drugs that cause physical dependence because of its durability of action whatever the percentage of dose. Methods of use and effects of heroin are similar to those of morphine, including mental and physical dependence. A lethargic, drowsy, "on-the-nod" physical state can be expected as a result of a dose of heroin. Inhaling it in powder form causes redness and rawness around the nostrils, owing to its acid content.

Normally, drug users do not start their needle habit by injecting directly into a vein, but begin with intradermal injections, or scratchings parallel to or just under the skin. "Skin popping," or subcutaneous injection, refers to entry of the needle beneath the skin surface. An intramuscular injection goes into the muscle and flesh of the body. Finally, intravenous injection, directly into the bloodstream by piercing the vein, will obtain the quickest reaction to the drug. (See Figure 10.2.) Injecting heroin leaves visible scars, or "track marks," that are usually found on the inner surface of the arms and elbows, although heroin users may inject into the body where needle marks may not be seen, such as between the toes or behind the knees. A syringe, bent spoon or bottle cap, eye or medicine dropper, matches, cotton, a nail or bent razor blade, a belt or tie, and needles are items that can be found in or as a part of the kit used to prepare and administer heroin. (See Figure 10.3.)

Heroin is big business, with a high return on invested capital, and

FIGURE 10.2 Drug addict "skin-popping"

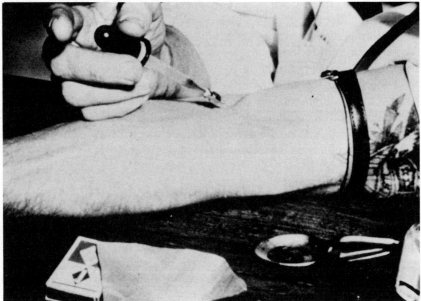

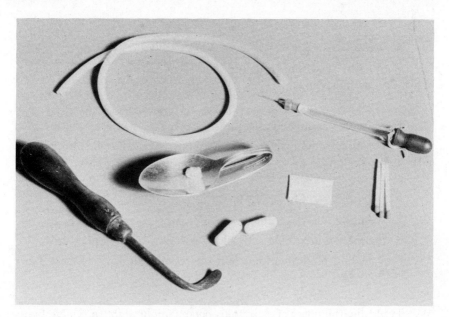

FIGURE 10.3   Narcotics equipment—"the works"

organized crime controls its trafficking and sale. The majority of the heroin entering the United States comes from poppy fields of Turkey, via Europe. The raw opium is smuggled out of Turkey and into Lebanon and other Mideastern countries, where it is converted into morphine base. From there, it finds its way to clandestine laboratories in France, where the final conversion into heroin is completed. Then, the heroin is smuggled to the North American continent for wholesaling, cutting, resale, and distribution down to the street level. Heroin is also received from Mexico, South America, and the Orient.

Heroin sold on the streets has diluted strength when compared to the heroin as it initially enters this country. For instance, heroin coming from Europe is in most cases over 90 percent pure. By the time it is "cut," or adulterated with milk, sugar, quinine, or mannite (a foreign laxative), the average street dose probably contains from 3 to 5 percent heroin. These items are used to cut heroin because they are like it in color, they provide bulk, and they readily dissolve in water with the heroin when the user prepares for a dose, or "fix."

There has been an increase nationally in drug-related deaths due to overdose (OD). Overdose is caused by taking a dose of a drug that contains a much higher percentage of the pure drug than the user's tolerance can withstand. If a heroin-dependent person is accustomed to getting the average street dose in his area—say, 5 percent—and he unknowingly purchases a dose containing 18 percent, the dramatic increase in potency may so severely depress his central nervous system that the resultant respiratory failure may cause death.

Symptoms of withdrawal and denial of substances causing morphine- and heroin-type dependency include:

1. Nervousness, anxiety, sleeplessness
2. Yawning, running eyes and nose, hoarseness, and perspiring
3. Enlargement of the pupils of the eyes, "gooseflesh," muscle twitching
4. Severe aches of back and legs, hot and cold "flashes"
5. Vomiting, diarrhea, and stomach cramps
6. Increase in breathing rate, blood pressure, and temperature
7. A feeling of desperation and obsessional desire to secure more of the drug

Typically, the onset of symptoms occurs about eight to twelve hours following the last dose. Symptoms increase in intensity, reach a peak after 36 to 72 hours, gradually diminish over the next five or ten days, and usually disappear entirely within ten to fourteen days. Weakness, insomnia, nervousness, and muscle aches and pains may persist for several weeks. It is possible to become physically dependent on more than one drug at the same time.

Codeine

Codeine, another derivative of opium, was discovered in 1832. It appears on the market in tablets and capsules, and in liquid form, and can be administered orally or by injection. It creates both mental and physical dependence in the abuser, but is milder in effect than morphine or heroin. It is prescribed extensively in medicine and is the base of many pain relievers and cough remedies. Codeine is normally used by drug abusers when more powerful opiates are not available. It is found in opium, but is produced from morphine.

Some Other Opium Derivatives

Dilaudid is a trademarked name for dihydromorphinone. It is derived from morphine, but is about five times as active biologically as morphine.

Paregoric, which is camphorated tincture of opium, has been used for many years to relieve pain. A drop or two, rubbed on the gums of a teething baby, can ease his pain considerably. Abuse, of course, consists of taking much larger amounts—a habit that would seem to be difficult, because of the unpleasant taste of the camphor.

Papaverine, named for *papaver*, the poppy genus, is a crystalline alkaloid, about 1 percent opium, that can be made synthetically from vanillin. It is properly used medically as a muscle relaxant.

Synthetic Opiates

Synthetic opiates can also create mental and physical dependence in the abuser, but are somewhat milder in their effect than morphine or heroin. The methods of administering and the physical implications of synthetic-opiate use are also similar to morphine and heroin.

Demerol is the commercial name for chemicals like isonipecaine, meperidine, and pethidine. Pethidine is popular in the practice of medicine at the present time.

Methadone is also known by several commercial names, such as Dolophine, Adanon, and amidone. A heroin-dependent person can be treated with doses of methadone to replace the heroin; subsequent doses of heroin will then have no effect on the abuser. Actually, this involves a transfer of one dependence for another, heroin for methadone, but methadone dependence is easier to control and treat medically.

Barbiturates

Barbiturates are a group of nonnarcotic drugs that are derived from barbituric acid. They are often referred to as "sleepers" or "downers," because of their depressant and sleep-producing effect on the central nervous system. The effects and symptoms of barbiturate use resemble the intoxication associated with alcohol, except that there is no odor on the breath. Constricted eye pupils may be apparent, as will slurred speech, drowsiness, lack of coordination, and disorientation.

Because barbiturates are among the most versatile of the depressants, they are used for treating epilepsy, high blood pressure, and mental disorders, and are administered before and during surgery. Alone or in combination with other drugs, they are used for many illnesses and situations requiring sedation, since they are capable of depressing or inhibiting the normal activity of nerves and muscles. Under medical supervision, barbiturates are safe and effective, but when they are abused and taken in large amounts, they can be very dangerous, because tolerance and physical and mental dependence develops.

Barbiturates come in tablets, powders, and capsules, and in liquid form, but are most often taken orally in tablets or capsules. Owing to the manner in which they are packaged and marketed, they appear in a variety of colors, and their slang names reflect this fact. For example:

1. Pentobarbital and Nembutal are called "yellow jackets."
2. Seconal and secobarbital are referred to as "reds," "red devils," "red birds," or "pinks."
3. Amobarbital tablets and capsules are called "blue birds," "blue devils," or "blue heavens."
4. Tuinal, a mixture of amobarbital and secobarbital, is called "rainbows," "double trouble," or "reds and blues."

Barbiturate withdrawal can be compared to the withdrawal symptoms previously discussed for narcotics; however, there are some important differences. Convulsions, which can be fatal, and failure of muscular coordination can occur during barbiturate withdrawal. Anyone suspected of suffering barbiturate withdrawal should receive medical help as soon as possible. Because of the threat of convulsions, many experts feel that barbiturate withdrawal is more dangerous than withdrawal from morphine or heroin.

Acute barbiturate intoxication now accounts for about 25 percent of all patients admitted to hospitals for treatment of acute poisoning. More deaths are caused by barbiturates used accidentally or on purpose than by any other drug. In addition, many deaths each year are attributed to drug users' not realizing the danger in mixing drugs—especially barbiturates and alcohol. The danger is in potentiation, the synergistic effect, which occurs when the combined action of two or more drugs is greater than the sum of the effects of each drug taken alone. The barbiturates and alcohol may be nonlethal doses individually, but in combination, their multiplication of effects can exceed body tolerance.

Most of the barbiturates are manufactured legitimately in massive quantities by pharmaceutical firms. They reach the illicit market by diversion and theft, by bogus prescriptions, or through poor control by the manufacturer.

Darvon

Darvon is a drug used by the armed forces and issued on prescription. The Darvon capsule contains a powder and a small pellet called propoxyphene, an analgesic, which abusers remove from the capsule and save in quantities to either take orally or prepare for injection. Darvon has, according to the manufacturer, an "addiction liability" similar to that of codeine.

Alcohol

Alcohol is a central nervous system depressant that, taken in sufficient amounts, impairs clear intellect, fine judgment, and muscular coordination. Studies are being conducted into the nature of alcohol dependency. Some experts contend that alcoholism is a psychological disorder, and others that it is addictive in much the same way as morphine.

The Stimulants

Stimulants excite or stimulate the central nervous system. They can be obtained from nature (for instance, from the coca plant) and can be synthesized by man.

Cocaine

Cocaine is obtained from the leaves of the coca plant, which is a squat bush that grows in the Andes Mountains in South America. An alkaloid, cocaine occurs naturally in the leaves in such concentrations that the natives obtain a stimulant effect from chewing the leaves.

When extracted and refined, cocaine is a white crystalline powder, similar to snow in appearance. It causes mental but not physical dependence. Tolerance does not develop, and abusers seldom increase their customary dose. Cocaine produces a sense of euphoria and a feeling of increased muscular strength, as well as increased heart rate and blood pressure.

Cocaine is usually sniffed, so that it contacts the mucous membranes in the nose, but it can also be injected. Some drug abusers use cocaine with other drugs, such as morphine or heroin. Combination shots of cocaine and heroin, cocaine and morphine, or all three are called "speedballs."

The aftereffects of cocaine include anxiety, restlessness, possible hallucination, and feelings of paranoia and depression. It causes a dilation of the pupils of the eyes, and users who sniff cocaine over a period of time develop running noses and sniff frequently.

Khat

Khat is a shrub that is found in East Africa, the Arabian peninsula, and Southeast Asia. The leaves, branches, and roots are used as a stimulant and euphoriant. Khat, also known as miraa and kat, is normally chewed or brewed into tea. Chronic abuse results in a loss of the sense of reality and a decline in physical and mental health.

Amphetamines

The amphetamines were discovered in 1927 in Los Angeles by George Allen, a pharmacologist. They are synthetic nonnarcotic dangerous drugs and are related chemically and pharmacologically to a group of compounds generally known as sympathomimetic amines, which act like adrenaline on the body. They have a marked stimulating effect on the central nervous system.

Amphetamines are widely used by such people as truck drivers and night watchmen, to keep them awake or increase alertness—although this is often a feeling rather than a measurable increase. They are also widely used medically to treat obesity, because they inhibit or suppress the appetite. For this reason, amphetamines are popular with women who are concerned with weight control, but many doctors no longer prescribe amphetamines for this purpose because of the abuse potential and the possible side effects. In addition to weight control, amphetamines are used medically to treat narcolepsy, a disease that results in involuntary attacks of sleep, and to counteract the effects of depressant drugs.

Amphetamines appear in capsule, tablet, or liquid form. They are most often taken orally but can be injected. Like the barbiturates, amphetamines are known by their street slang names: "dexies" (dexedrine), "bennies" (benzedrine), "uppers," "copilots," and "pep pills."

It is paradoxical that these drugs, although stimulants, tend to calm hyperactive, noisy, and aggressive children and channel them toward normal behavior patterns. Some school systems have been using them to treat children with learning difficulties, but this is an area of much controversy at the present time. There are possible side effects to these drugs that many people feel are too risky to chance with young people.

Amphetamine abuse creates mental but not physical dependence. As abusers develop body tolerance, frequency of use and dosage increase. Abused in high doses, amphetamines may lead to a rise in blood pressure, palpitations, dry mouth, sweating, headache, pallor (from lack of eating or improper diet), and dilation of eye pupils. Abuse can also lead to erratic behavior and serious mental disturbance. The severity of psychotoxic effects increases when the drugs are injected intravenously.

Some doctors feel that amphetamine abuse can lead to pronounced personality change. Additionally, there is a possibility that excessive amounts of amphetamines can lead to permanent organic damage to the brain. Research is continuing to explore this effect in view of studies already completed in this area.

The Food and Drug Administration has appealed to doctors to stop prescribing amphetamines except for three disorders—short-term treatment of obesity, narcolepsy, and hyperactivity in children. Manufacturers have been asked to reduce their promotion and sale of the pills.

Methamphetamine (Speed)

Methamphetamine, another form of amphetamine, is called "speed" because of its rapid stimulation of the central nervous system. The term "speed" is also used for other stimulants and amphetamine-like substances, but is most related to methamphetamine because of its stronger action. Speed use has increased in this country, and in many areas, it has replaced LSD as the chosen drug of abuse. Speed appears in capsule, tablet, or liquid form.

Normal therapeutic doses of amphetamines may be from 5 to 15 milligrams, but methamphetamine abusers, "meth heads" or "speed freaks," inject many times that amount—perhaps hundreds or thousands of milligrams in a single dose—owing to the tolerance they have developed. As much as a 15,000-mg dose has been recorded as injected in one case during a 24-hour period.

With some people, the fascination with speed lies in the initial effect of the drug. Upon injection, a "rush" or intense feeling of euphoria often results that, like that associated with heroin, has been compared to sexual orgasm. For this reason, speed is supposed to have aphrodisiac qualities for some users; in others, impaired sexual potency has been reported. Usually, as the abuse progresses, there is a reduction of sexual interest.

The "speed run" is a prolonged period of time, over a few days or a week, when the abuser injects methamphetamine as often as is necessary to feel the desired effects. During this period, the "speed freak" usually does not eat or sleep. Initially, he may feel a sense of paranoia; then he becomes extremely suspicious of others, hallucinates, and is overactive. The combination of suspicion, hyperactivity, impulsiveness, and irrational thinking has often resulted in aggressive and destructive behavior. After a speed run, the abuser "crashes" and is "amped out" in a state of depression and exhaustion. A period of sleep usually follows, lasting from 24 to 48 hours. Upon awakening, the speed user feels depressed and miserable.

There are many aftereffects of speed use that can cause physical and mental harm: malnutrition; prolonged psychotic states; skin infections and tooth decay; heart trouble and convulsion; hepatitis from dirty needles and solutions; liver damage; suspected brain damage manifested by apathy, memory gaps, confusion, and disorientation; and personality disturbance.

Most methamphetamine abusers evince immaturity and try to block their limitations by using speed as a crutch. They cannot handle or tolerate frustration, and want immediate satisfaction of desires. It has been written that speed is a total ego trip, and that the "speed freak" has little time for anything or anybody other than his love affair with speed.

A University of Southern California, Los Angeles, research team recently predicted that large numbers of young people will soon require long-term care for a frequently fatal disease, necrotizing angitis, which is caused by heavy abuse of methamphetamines, or a combination of amphetamines with other drugs. The disease, for which there is no known effective cure, causes inflammation of the arteries and leads to failure of major organs, such as the liver, kidneys, pancreas, and small intestines. It should be remembered that the liver is the principal detoxifying organ in the body, and if its function is destroyed, so is the life that it supports.

Amphetamine Equivalents

These are amphetamine-like compounds that have the same approximate pharmacological properties as amphetamines, and that are sometimes incorrectly called amphetamines. Two amphetamine equivalents that are widely used drugs of abuse, not only in this country but elsewhere in the world, are Preludin (phenmetrazine hydrochloride) and Ritalin (methylphenidate hydrochloride). These are used by doctors as antidepressants and in the treatment of narcolepsy.

The Hallucinogens

The hallucinogens are the psychotomimetic (producing reactions that are imitative of psychosis) substances that alter the user's state of consciousness. They encompass a broad spectrum of drugs and compounds, some of which are mild in nature and reaction, while others cause devastating results. As with other forms of drug abuse, many reasons are given for the abuse of hallucinogens. Frequently, the reason seems to be a belief that one can use them to obtain a heightened state of awareness, in order to seek and truly understand the meaning of truth, love, knowledge, and values. Whatever the reason, the effects dramatically change the life style of many abusers.

Mescaline and Peyote

Mescaline is derived from the buttons of the peyote cactus, which grows in Central America and the southwestern United States. It is currently popular among the drug culture because it is thought to be a smooth and safe hallucinogen. The use of mescaline and the chewing of peyote buttons has been practiced for centuries by various Indian tribes, both as a medicine and as a part of religious ceremonies. It is still used by Indians who are members of the Native American Church, to induce mystical visions.

In order to facilitate ingestion, peyote buttons are ground into powder and taken orally. Mescaline is available on the illicit market as a crystalline powder in capsules, or as a liquid in ampoules or vials. Because of its bitter taste, the drug is injected or eaten with food or beverage. Mescaline produces illusions and hallucinations lasting from five to twelve hours, and creates psychological dependence.

Psilocybin and Psilocyn

These drugs are obtained from mushrooms that grow in Mexico. Like mescaline, they have been used by Indians for centuries. The drug experience lasts about six hours. Neither drug produces physical dependence, but users have been known to develop a tolerance to them, requiring larger doses to achieve the stimulation desired.

DMT (Dimethyltryptamine)

DMT is one of the tryptamine series and is found in the seeds of certain plants native to the West Indies and to South America. The powdered seeds have long been used as snuff, called "cohoba." This drug has been produced synthetically by clandestine laboratories in the United States for the illicit market. DMT is usually not taken orally. Its vapor is inhaled from the smoke given off by burning the ground seeds or synthetic powder mixed with tobacco, parsley leaves, or even marihuana. It can also be injected. "Trips" are short, but mental dependence may result. DMT produces effects similar to those of LSD, but much larger quantities are needed.

Bufotenine

Bufotenine is related chemically to DMT, but is extracted from the dried glandular secretions of certain toads. It may be taken by injection or used like snuff. Its symptoms are immediate and severe, since its use results in visual disturbances and alteration of time and distance perceptions. It also seriously affects the blood pressure of the user.

DET (Diethyltryptamine)

This drug is also related chemically to DMT, but has not yet been found in plant life. It can easily be produced in a laboratory. Injected doses of 50 to 60 milligrams cause visual distortions, dizziness, and a vague sense of time. The experience may last from two to three hours.

DOM

More commonly known as STP, DOM appeared on the drug scene early in 1967. Underground newspapers promoted its use, claiming that it was stronger than LSD. However, it has been found to be about 1/15 as strong as LSD in the average dose. "STP" in the drug culture stands for serenity, tranquility, and peace. A study conducted at Johns Hopkins University found that mild doses lead to euphoria, and stronger doses to genuine hallucinations; effects have been noted to last up to ten hours. STP is not found in nature, but is synthesized in laboratories and formed into tablets slightly larger than LSD capsules or tablets. One STP form is called the "magic pumpkin seed" because of its long, yellow appearance.

MDA (Methylene Dioxyamphetamine)

MDA is a relatively new hallucinogen. It has appeared in powder and tablet form. Information as to its comparative strength is still uncertain, but it appears to be weaker in reaction than LSD.

PCP (Phencyclidine)

Sometimes called the "peace pill," PCP is a compound that has been used as a veterinary anesthetic. It has been seen in colored tablets, mixed with LSD or alone, and as a white powder in capsules, which can be smoked on parsley. Sometimes the latter form is called DOA—"dust of angels." PCP has been used by street suppliers of drugs as a substitute for LSD, synthetic THC, and other drugs. The naive and unsuspecting buyer sometimes does not realize the difference; as long as he receives some sort of physical and mental reaction, he may conclude that he had a weak form of the real thing.

DOET

This is an ethyl component in a new mescaline analog that was developed as a treatment for chronic depression. It allegedly does this by awakening or sharpening perceptions through some sensory stimulation.

LSD (D-Lysergic Acid Diethylamide)

LSD was synthesized by Albert Hoffman at the Sandoz Laboratories in Switzerland in 1938. It is derived from the lysergic acid present in ergot, a fungus that grows on rye. LSD was brought to the United States in 1948 by two doctors for experimental treatment in mental health; it failed in both alcoholic and mental cases. Research continued and, by 1960, illegal production, distribution, and use of LSD began to increase. Use has declined recently in some areas as drug abusers have recognized its dangers and have moved on to methamphetamines. At the present time, there is no legitimate manufacturer of LSD in the United States.

LSD comes in two forms: dextro and levo. The dextro (D) form is the active form; it is almost 100 times stronger than the levo (L) form, which causes no reaction.

LSD is a tasteless, odorless, colorless liquid in its pure state and is normally taken orally. On the illicit market, it can be found as a tablet, crystalline powder in various capsules, or in liquid form in ampoules. It is often impregnated in sugar cubes, cookies, or crackers, and can be put on the back of postage stamps, or on letter paper to be eaten by the receiver.

This is the most potent hallucinogen known. One ounce contains a sufficient amount for about 300,000 doses or LSD experiences. This LSD on the tip of a pin would be sufficient for several "trips." The average dose in illicit LSD "caps" contains about 100 micrograms.

LSD primarily affects the central nervous system by producing changes in mood and behavior. It may also dilate eye pupils, cause tremor, elevate temperature and blood pressure, and produce hyperactive reflexes in the user. Tolerance to the behavioral effects quickly develops with several days of continued use. As with the other hallucinogens, physical dependence may not occur, but minor mental dependence may develop.

Other manifestations of LSD use include hallucinations, panic or paranoia, extreme anxiety, mental depression with suicidal thoughts or attempts, and "release" from reality to the point that the user does not know who or what he is. These are unpredictable reactions that may not be experienced by all users.

Flashback, a bizarre aftereffect of the LSD experience, is the reoccurrence of the LSD trip at some unknown future time. It can occur after one LSD trip or after 50. Some people who have used LSD never experience flashback; others experience it more than once. Flashbacks usually occur with the same intensity as the original LSD experience, but not necessarily for the full duration. There is no clear evidence to explain flashback or what causes it, since LSD cannot be traced in the body. It

FIGURE 10.4
Marihuana leaf

FIGURE 10.5 Marihuana leaf, seen through a microscope

is possible that the LSD molecules break down, attach themselves to other chemicals already in the body, and lie dormant until something at some future time triggers their release to cause the flashback; or that it could be triggered by some sensory or auditory stimuli that closely resemble physical or environmental conditions during the original experience.

Because LSD is such an elusive chemical in the body, not all its effects are known. For instance, there is concern about the possible occurrence of chromosome damage to its users, and about both chromosome damage and anatomical damage to their unborn children. A 1970 report by George Washington University, the first long-term study of its kind, might indicate where further research is needed and the results that can be expected by future studies. The study covered a period of two years and followed 112 women through 127 pregnancies as nearly as possible from the time of conception. All subjects in the control group had used LSD prior to, during, or after conception. As a group, these women experienced 18 times the rate of serious birth defects in their offspring that was found in the general population. The rate of spontaneous abortion was nearly double that of the general population.

Marihuana (Cannabis Sativa L.)

Of the illicit drugs of abuse, marihuana is the greatest problem. Only alcohol, which is legal, exceeds marihuana as a drug-abuse problem.

Marihuana (Figure 10.4) is a sturdy hemp plant that grows like a common weed almost anywhere in the world with little or no care. It contains an active chemical ingredient in the resin of the plant called tetrahydrocannabinol (THC), which was first isolated in 1940 and has been synthesized since 1966. THC is the chemical that produces the euphoric effects; it is found in highest concentration in the flowering leaves and tops of the female marihuana plant. Marihuana plants from the tropics and warm climates have a higher THC content that those from cooler climates. For this reason, marihuana from Mexico or Southeast Asia is preferred to our own domestic variety.

It is relatively easy to identify the plant. The compound leaves are the key; each of the larger ones consists of five to eleven leaflets, but generally seven. Each leaflet is covered with small hairs, and has notched edges and pronounced veins. (See Figure 10.5.) The upper side of the leaf is deep green in color and the lower side is lighter green.

Marihuana is usually smoked in cigarettes or pipes for the fastest effect (see Figures 10.6 and 10.7), although it can be eaten, mixed with tea or other liquids, or mixed into or baked in food. The leaves must first

FIGURE 10.6
Marihuana cigarette

FIGURE 10.7
Hollowed-out cartridge, used as marihuana cigarette holder

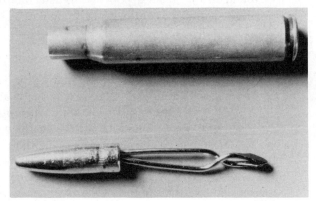

be dried and then broken up or crushed. Some marihuana is roughly cut and contains stalks and seeds. Better grades are chopped and have most of the seeds and debris removed, producing a more desirable cigarette; it is thus said to be "manicured." Manicured marihuana is the finest grade for sale and commands the best price, owing to the high concentration of leaves, which contain the THC. Marihuana in its most concentrated form is called hashish.

Marihuana abuse on a regular basis can and does lead to mental dependency. Reactions to its use are unpredictable, because of the questionable THC content. Generally, however, there is a lack of overt manifestations; a person can mask the effects of a marihuana "high" much more easily than can a person who is intoxicated with barbiturates or alcohol.

A distortion of time is one of the effects. What seems, in the mind of a marihuana smoker, a period of five minutes may actually be only two. Marihuana smokers have difficulty remembering things that have occurred in the immediate past (say, in terms of seconds). This can be critical if they are functioning in a stress situation.

Other effects are alterations in depth perception, powers of concentration, peripheral vision, distance perception, emotions, and motivation, and marked paranoia may occur. In addition, the user may experience dizziness; dry mouth; dilated eye pupils and burning, bloodshot eyes; urinary frequency; diarrhea; nausea; hunger (especially for sweets, owing to a lowering of the blood-sugar level in the body from chronic abuse); and throat irritation.

One of the primary concerns about marihuana is its potential effect as a reality-distorting agent and the impact of that effect on the psychological development of the user, especially the adolescent. Normal adolescence is a time of great psychological turmoil. Patterns of coping with reality that are developed during the teen-age period are significant in determining adult behavior. Persistent use of an agent such as marihuana, which serves to ward off reality during this critical development period, is likely to seriously compromise the youngster's future ability to make adequate adjustment to society. Characteristic personality changes induced among impressionable young people from the regular use of marihuana include apathy, loss of effectiveness, and a diminished capacity or willingness to carry out complex long-term plans, to endure frustration, to concentrate for long periods, to follow routines, or to successfully master new material in the learning process.

Marihuana tends to lessen inhibitions and increase suggestibility. This explains why a person under the influence of marihuana may engage in activities he would not ordinarily consider.

Others

The deliriants of the belladonna family have been used since ancient times to produce visions, mental distortions, and confusional states. They include belladonna, datura, henbane, and others. At the present time, only two members of this group have come to the attention of authorities as items of abuse; these are:

1. Jimson weed, which grows wild in many parts of the United States.

2. Asthmador, which is a patent medicine for asthma that contains a mixture of belladonna and datura strammonium. When abused, it causes delirium.

Hawaiian baby wood rose, nutmeg, and morning glory seeds can provide delirium when taken in large amounts. Some of the seeds contain lysergic acid amide.

Propellants for aerosol sprays, such as freon and nitrous oxide, can cause physical damage if inhaled, by preventing oxygen from getting to the lungs.

The Volatile Chemicals

The volatile solvents include model-airplane glue, lacquer thinner, gasoline, fingernail-polish remover, and lighter fluid. These substances contain xylol, cresol, naphtha, benzol, tetraethyl lead, and other chemicals that cause acute or chronic damage to the body by attacking the oxygen level.

The primary method of abuse is by inhalation. Glue manufacturers recently started adding mustard to their glue formulas, in order to induce nausea and curb glue sniffing.

INVESTIGATIVE CONSIDERATIONS

Most of the techniques used in dangerous-drug and narcotics investigation are the same as those used during other cases; however, there are a few variations with which the investigator should be familiar. This section treats the background of the availability and acquisition of the drug by the offender through its identification by the investigator.

Acquisition of Narcotics, Dangerous Drugs, or Marihuana

One of the means of tracing illicit drug sources is to determine supply outlets. This can be accomplished in many ways—for instance, through review of completed reports of investigation, informant information, or surveillance and direct efforts to purchase drugs. In the case of direct purchase, it is desirable to make more than one purchase from a peddler, if possible. This procedure gives agents more opportunity to locate the peddler's cache of narcotics and/or his source of supply. It also serves to identify other customers and helps establish that the peddler is a regular participant in the illegal narcotics traffic and not a one-time or opportune offender. As the sources of supply and customers of a peddler are identified, the possibility of formulating a conspiracy involving others increases.

If an arrest is going to be made immediately after a purchase, it may be desirable to dust the currency used with fluorescent powder. In addition, the serial numbers of money used to purchase narcotics should always be recorded for subsequent comparison with any money recovered from the defendants. It may show an unsuspected link between known and unknown narcotics traffickers and may be a further indication of a conspiracy. Recovered money used for the purchase of narcotics should be treated as evidence.

If a peddler will sell only to an informant and not to an agent, the informant must be searched prior to the sale for money or narcotics. Any money found on his person should be removed and then returned upon completion of the purchase. He should be searched again immediately after the sale to ensure his integrity. Between the two searches, the informant must be kept under constant surveillance in order that his testimony can be corroborated. Evidence obtained through this method, "informant buys," is admissible in trials, but the testimony of the informant may be required.

The district attorney should be consulted concerning the adequacy of a surveillance during an informant buy when the informant is lost from view, as when he enters a building where the actual sale is transacted. All exits of the building should be kept under surveillance while the informant is inside. An electronic transmitting device, which can be monitored and recorded by the surveillance agents, should be utilized during a purchase of narcotics. It is desirable to have the undercover agent or informant equipped with a miniature recorder, such as a Minifon, for a reserve in case the transmitting device fails to function. Sales should be photographed.

Apprehension of a Violator

With some notable exceptions, the precautions taken and procedures used for the apprehension of any dangerous criminal apply equally to the drug violator. The seriousness of drug laws is best evidenced by the penalties invoked for their violation. Federal laws provide for a five-year mandatory minimum sentence, in many cases without possibility of probation or parole, for first offenders, and a ten-year mandatory minimum sentence without possibility of probation or parole for second or subsequent convictions. The death penalty may be imposed under the federal laws for the selling of narcotic drugs to a person under the age of 18.

The apprehension of a narcotics violator should be done as unobtrusively as possible, in order to prevent the knowledge of it from reaching his collaborators. In many cases, the investigation may benefit from their not knowing.

When approaching a drug suspect, the investigator should particularly observe the suspect's hands. He may attempt to dispose of the drugs in his possession by dropping, throwing, flushing, eating, or otherwise disposing of the contraband. If the drug should be found on the ground some distance away from the suspect, it may be extremely difficult if not impossible to connect him with it. Also, since addicts can be dangerous, unpredictable, and uncooperative, attention should be given to the possibility that they have weapons in their possession.

As soon as possible after apprehension, the suspect should be required to place his hands directly in the air or behind his head to preclude further disposal of evidence. The investigator must be sure that drugs he takes from the custody of a person suspected of narcotics abuse are not prescribed for the health and well-being of the suspect. If the suspect alleges that certain seized drugs are necessary for his health, a medical opinion must be obtained prior to allowing any administration of the supposed medication. The mere possession of a prescription for the drug should not by itself be used as justification for not seeking the opinion of a physician.

Immediately after apprehension, or as soon as possible thereafter, a thorough search should be made of the suspect, his clothing, and the area within his immediate control. The small packages in which illegal drugs are usually dispensed can be easily hidden in a very small space. The possession of even the minutest of particles may be sufficient to establish a conviction. The following are places where narcotics have been found on the person of narcotics violators: inside hatband, coat lining, shoulder pads, sleeves, pants, waistband, cuffs, seams, tie, shoes, cigarettes, lighter, pen or pencil, chewing-gum package, hair (including pubic hair), and body orifices. On teenagers, boots and false heels are favorite hiding areas. The investigator must be alert to the presence of all tablets, capsules, small pieces of paper, and liquids, as well as the more conspicuous types of equipment such as syringes, needles, medicine droppers, and bent and/or discolored spoons.

If it is necessary for a narcotics violator to be apprehended in a building, an agent should attempt to position himself between the suspect and the bathroom or any sink. Toilets and the drains of sinks, basins, bathtubs, and showers are favorite hiding places for narcotics, since they offer a quick and easy self-destruction capability. Another favorite method of operation is to hang the drugs out of a window by a string held in place by the closed window. Merely opening the window releases the cache and permits it to fall to the ground. Unless the drugs are seen falling from the window, the possibility of connecting the suspect to the cache is remote.

Religious artifacts should be checked thoroughly when a search is made for hidden drugs, since they are often used as hiding places in hopes that searching agents will overlook them. Drugs have also been found inside tires, on the roofs of glove compartments, and under the dashboards of automobiles; in firearms; inside toilet tanks; in chimneys; and in hollowed-out furniture. The variety of hiding places for narcotics is limited only by the ingenuity of the violator.

Title 49, United States Code, Sections 781 and 782 provide for the seizure and forfeiture of any "vessel, vehicle, or aircraft" used to "facilitate the transportation, carriage, conveyance, concealment, receipt, possession, purchase, sale, barter, exchange, or giving away of any contraband article." The definition of "contraband article" in this law includes illegally sold or obtained narcotic drugs or marihuana.

Every investigator should be familiar with the elements of proof for a conspiracy, so that he will recognize the crime when the elements are present. Narcotics violations are often conspiratorial by nature and are particularly fertile fields for aggressive conspiracy investigations. Efforts to develop proof of a conspiracy through the interrogation of suspects should always be made, since many of the drug violators in higher echelons seldom come in direct contact with narcotics. In the development of evidence to prove a conspiracy, some valuable tools are telephone company records, testimony of co-conspirators, bank records in the tracing of monies, records of gasoline credit-card purchases, and credit records.

Disposition of a Prisoner

Investigators should be very careful in questioning heroin addicts and should seek medical help when withdrawal commences. Failure to use these precautions could easily be considered duress by the courts, and any information obtained under these circumstances could be ruled inadmissible. It should be recognized, however, that the addict will frequently feign an excessive degree of suffering in order to elicit sympathy and favored treatment. Since true withdrawal symptoms create real pain, the prisoner should be taken to a medical facility for proper treatment. If it is unknown whether the prisoner is a true addict or one who only occasionally uses a drug, a medical officer may be able to determine the degree of narcotic addiction through the use of the Nalline test, although this test, because of tolerance considerations in determining the dosage, is not an infallible one.

Processing of Drug Evidence

The processing of drug evidence is carried out in the same manner as that of other items of evidentiary material. Meticulous care in maintaining the legal chain of custody is extremely vital, in order to introduce drugs into evidence at the time of trial. The number of people who come into possession of drug evidence should be kept to an absolute minimum.

Drug evidence is usually found in minute quantities and often in small containers. They should be placed in a suitable outer container as soon as possible, and the container marked with the investigator's initials and the date the evidence was obtained. Marihuana cigarettes can be marked on the cigarette paper. Without chemical analysis by a competent chemist, pills, capsules, powders, and vegetable matter cannot be positively identified as drugs. In the case of substances suspected of being drugs, the agent must refer to the evidence by its physical appearance: for example, "a white crystalline powder suspected of being codeine," or "vegetable matter suspected of being marihuana." Exact-weight statements should be avoided. The amount may be referred to as "approximately 1 cup of vegetable matter," "¼ teaspoon of powder," "24 tablets."

Drugs and drug evidence should be stored in a safe or security container inside the evidence room for proper security, and should be kept there at all times except when required during the course of the investigation or trial.

Identification of Illicit Laboratories by the Reagents Present

There are many chemicals, such as solvents (ether, alcohol, chloroform, etc.) and reducing agents (lithium aluminum hydride) that are utilized in the synthesis of many or all of the drugs mentioned. The only reagents included in the table on page 231 are those used specifically in the manufacture of a single drug or group of drugs.

The presence of any of the following chemicals in an illicit laboratory is a strong indication that the drug under which it is listed is the one being synthesized:

LSD

Ergotamine tartrate
Lysergic acid
Nitrogen
Dimethylformamide
Sulfur trioxide
Acetonitrile
Trifloracetic anhydride

Thirty milligrams of D-lysergic acid can be made with 100 milligrams of ergotamine in 12 hrs. In 8 more hours, 40 milligrams of LSD can be made.

Mescaline

3, 4, 5 trimethoxyphenylacetonitrile
3, 4, 5 trimethoxybenzoic acid
3, 4, 5 trimethoxybenzyl chloride
3, 4, 5 trimethoxybenzyl alcohol
(Explosive—Lithium Aluminum Hydride)

Three hours to convert to Mescaline.

DMT

Indole
Oxalyl Chloride
Tetrahydrofuran

Six to 9 hours to convert to DMT.

Amphetamine Sulfate

Phenylacetone (phenyl 2-propanone)
Formamide
Hydroxyl amine

Methamphetamine (called "crystal" on the street)

Phenylacetone (phenyl 2-propanone)
Ephedrine
Zinc or tinfoil

Field Identification Tests for Dangerous Drugs

Police crime laboratories are staffed with qualified chemists and have excellent equipment with which to identify suspected material. (See Figure 10.8.) They utilize advanced techniques and sophisticated materials for chemical analysis. Laboratory examinations are precise, expert, and, in most cases, capable of positive proof of drug existence. The

FIGURE 10.8
Measuring weight of unknown liquid

personnel of these laboratories are prepared to testify in court as to the chemical analysis of suspected material.

Field tests are preliminary and can indicate only that a drug may be present; therefore, the chemical field tests for drugs listed below are precursory and presumptive only. They can be used by investigators to screen many of the commonly used drugs offered for sale by illegal sources. Any drug that will be used as evidence, however, must be positively identified by a qualified chemist, utilizing approved procedures in an adequately equipped laboratory. If the suspected material is minute, field tests should not be attempted and all of the substance should be forwarded to the laboratory for analysis. In many cases, the color reactions produced by field tests are only indicative that the suspected sample is a drug product. Finally, the testing of a suspected material through sampling should never be permitted or practiced. There is a great danger that poisons might be introduced in a drug or that the material may, in fact, be a poison.

Field Test for Amphetamines

This field test for identifying amphetamines is useful in screening out caffein, vitamins, saccharin, or other substitute tablets sold in lieu of amphetamines.

Test Material. The test material consists of two or three drops of Marquis reagent (2 drops of 37 percent formaldehyde in 3 ml concentrated sulfuric acid) in a small glass ampoule. The Furguson Company, 814 Ridgeley St., Baltimore, Md., produces kits consisting of twelve ampoules of Marquis reagent packed in a small container that can be carried in a coat pocket.

Caution: Marquis reagent contains concentrated sulfuric acid. Care must be taken not to get it on bare skin or clothing. Before breaking, the ampoule should be held in a vertical position to drain the reagent to one end. Dispose of the broken ampoule carefully.

Test Procedure. Break the ampoule at the scored center and place one or two drops of the reagent on the whole or crushed tablet. This should be done on a glass ashtray, inverted tumbler, or other surface that will not be affected by the acid. White, pink, yellow, peach, or green amphetamine tablets react with the reagent to give a red-orange color, turning to reddish and then dark brown within one or two minutes. The speed with which the color is formed appears to depend upon the hardness of the table. The red-orange color forms immediately on some tablets; others require 10 to 20 seconds. Therefore, the critical period of color differentiation for amphetamines is within the first 20 seconds.

The peach-colored caffein tablets produce a color that might cause some confusion. The difference between the color formed by this tablet and that formed by a peach-colored amphetamine tablet seems to be more obvious if the tablets are crushed before the reagent is applied. Once the difference is seen, there should be no trouble in distinguishing one from the other.

As indicated in the table below, some examples of materials that give the same color change as amphetamine are the phenyl tertiary butylamine HCI tablets and the wyamine sulfate tablets. Both are similar chemically to amphetamines.

Marquis reagent can also be used as a field test for detecting narcotics. Opium derivative alkaloids yield a purple color. Morphine gives an intense purple-red, changing to violet, and heroin gives a reddish-purple color.

Listed below are the color reactions of the materials of interest.

1. Amphetamine powder and tablets	Red-orange immediately; reddish-brown to dark brown within a couple of mins.
2. Caffein powder and tablets	No color
3. Methamphetamine tablets	Red-orange to reddish-brown and then to dark brown in 1 or 2 mins.
4. Phenyl tertiary butylamine HCI tablets	Same color change as amphetamine tablets
5. Wyamine sulfate (N methylphenyl tertiary butylamine sulfate tablets)	Same color change as amphetamine tablets
6. Mescaline Sulfate	Bright orange-red, which does not change

Field Test for Barbiturates

For the tentative identification of barbiturates, the Zwikker Test is used. An anhydrous methanol solution of the barbiturate, upon the addition of several drops of cobalt chloride in methanol solution, gives a bluish color, which changes to dark blue upon being alkalized with a 5 percent isopropylamine in methanol.

Test Materials. A compact kit for administration of the Zwikker Test is manufactured by the Atkinson Laboratory, Inc., 3031 Fierro St., Los Angeles, Cal. The kit consists of a small plastic box, containing a small porcelain spot plate and three solutions in plastic dropping bottles:

1. Solution No. 1—Anhydrous methanol
2. Solution No. 2—Cobalt chloride (cobaltous chloride) dissolved in methanol
3. Solution No. 3—5% isopropylamine in methanol

Caution: These solutions are volatile and flammable. They should be kept stoppered!

Test Procedure. Place part of the contents of the suspected capsule, powder, or crushed tablet into spot-tester (enough to cover the letter O on a typewriter key). Add two drops of Solution No. 1 to fragments on spot-tester (fragments should dissolve). Add two drops of Solution No. 2 (this may produce a violet or blue color). Then add two drops of Solution No. 3. If the color deepens to a darker violet or blue, it indicates presumptive presence of a barbiturate.

Caution: Do not let the dropper touch the sample, as the solution will become contaminated. Wash and dry the spot-tester plate after each test.

Field Test for Mescaline

Test Material. For the tentative identification of mescaline, Marquis reagent is used. (See the test material under the field test for amphetamines.)

Test Procedure. Mescaline is usually encountered as the sulfate salt in 500-mg gelatine capsules. Place a small portion of the material contained in the capsule in a spot-test plate, glass ashtray, or inverted tumbler. Break the ampoule containing the reagent at the scored center and place one or two drops of the reagent on the material to be tested. Mescaline reacts to give a bright orange-red color.

Field Test for Use of Opiates (Nalline Test)

Nalline (N-allylnormorphine), an opiate when used alone, has some of the effects of morphine and other opiates. It is also a useful antidote to poisoning by opiates or methadone, but its most common use is in the diagnosis of addiction to these drugs.

Small doses of Nalline precipitate withdrawal symptoms in active addicts to these drugs and produce signs of use in those who have taken only a few doses. The symptoms of abrupt withdrawal appear almost instantaneously after intravenous administration, and in a matter of minutes after subcutaneous administration.

An authorization and release form should first be obtained from the subject, so that the test is on a strictly voluntary basis. Suspected users of drugs should be given a careful examination; if any abnormalities are noted, Nalline should not be administered. After thousands of tests, nothing more distressing than an occasional nauseated condition has been observed as a result of the administration of a minimal amount (3 mg) of Nalline. (If a person is exhibiting signs of withdrawal symptoms, the test should not be given, for it would be pointless and would simply hasten and intensify the discomfort.) In the event an addict is injected with the drug and does become ill, the administering doctor has available an antidote, such as morphine sulfate or an equivalent opiate, which counteracts the reaction and makes the addict comfortable again.

Field Test for Marihuana

Duquenois Test. The test is conducted as follows:

1. Extract 30 to 100 milligrams of the sample with 15 to 20 ml of petroleum ether.

2. Filter and evaporate to dryness.

3. Add 2 ml of Duquenois reagent and stir to bring the residue into solution.

4. Add 2 ml of concentrated hydrocholoric acid and stir.

5. Allow to stand ten minutes. A color will develop.

6. Transfer this colored solution to a test tube and shake with 1 to 2 ml of chloroform.

7. A violet or indigo-violet color will be transferred to the chloroform layer if marihuana is present.

Rapid Modified Duquenois Test. To a small portion of the suspected material, add 1 to 2 ml of Duquenois reagent, and shake for one minute. Add 1 ml of concentrated hydrocholoric acid. If the substance is marihuana, color will be observed first as a transitory green, changing to a slate gray and finally to indigo-violet. If the marihuana is fresh, the indigo-violet color will develop rapidly. Transfer this material to a test tube and add 1 to 2 ml of chloroform. The chloroform will form a bead or layer within the first reagent. The indigo-violet color will be transferred to the chloroform layer only if marihuana is present.

Thin-Layer Chromatography. Thin-layer chromatography field kits are sold commercially. The test is conducted by placing an extract of the questioned material on a thin-layer sheet and placing it in a developing jar. Separation of the major constituents of marihuana results.

Field Test for Opium Derivatives

Prepare Mayer's reagent: Dissolve 13.55 grams of mercuric chloride and 50 grams of potassium iodine in water to make 1 liter. Also prepare Marquis reagent: Add 1 ml of 40 percent formaldehyde solution to 20 ml of concentrated sulfuric acid.

Dissolve a small amount of the suspected sample in a few drops of Mayer's reagent. The appearance of a white milky precipitate indicates the presence of an alkaloid.

Add a few drops of Marquis reagent to a suspected opium derivative. If the derivative is present, it will show an intense purple-violet color immediately, which becomes more intense on standing.

Field Test for Methadone

Dissolve 1 gram of cobalt acetate, nitrate, or chloride and 1.5 grams of potassium thiocyanate in 90 ml of water and 10 ml of glacial acetic acid. Dissolve the sample in a minimum amount of water and filter it, if possible. Add 2 to 3 drops of the reagent. Shake about 1 minute. A blue precipitate indicates the presence of methadone.

Transmittal to Laboratory

When suspected drugs are to be shipped to the laboratory for analysis, care should be exercised to ensure that the packaging container is completely sealed, to avoid spillage in transit. Tablets and capsules should be packed in sterile cotton and placed in a suitable container.

Liaison with the DEA

Investigations involving narcotics or marihuana must be coordinated with the nearest office of the Drug Enforcement Administration (DEA). The DEA will request the available information regarding the offense and determine whether it will assume or relinquish investigative jurisdiction.

The DEA is interested in obtaining drugs that have come from the illicit market and that originate in clandestine laboratories. Narcotics that have served their purpose as evidence should be disposed of as soon as possible, and disposition instructions to the evidence custodian may call for releasing them to the DEA. If so, a form should be used to preserve a record of chain of custody after their release.

Bibliography

ALEXANDER, HAROLD L., *Classifying Palmprints*. Springfield, Ill.: Charles C Thomas, Publisher, 1973.

ARTHER, RICHARD O., *The Scientific Investigator*. Springfield, Ill.: Charles C Thomas, Publisher, 1973.

BATES, BILLY PRIOR, *Identification System for Questioned Documents*. Springfield, Ill.: Charles C Thomas, Publisher, 1970.

———, *Typewriting Identification*. Springfield, Ill.: Charles C Thomas, Publisher, 1971.

CROWN, DAVID A., *The Forensic Examination of Paints and Pigments*. Springfield, Ill.: Charles C Thomas, Publisher, 1968.

DAVIS, JOHN EDMUND, *An Introduction to Tool Marks, Firearms and the Striagraph*. Springfield, Ill.: Charles C Thomas, Publisher, 1958.

DIENSTEIN, WILLIAM, *Technics for the Crime Investigator*. Springfield, Ill.: Charles C Thomas, Publisher, 1974.

FIELD, ANNITA T., *Fingerprint Handbook.* Springfield, Ill.: Charles C Thomas, Publisher, 1971.

HALL, JAY CAMERON, *Inside the Crime Lab.* Englewood Cliffs, N.J.: Prentice-Hall, Inc., 1973.

HEFFRON, FLOYD N., *Evidence for the Patrolman.* Springfield, Ill.: Charles C Thomas, Publisher, 1972.

HORGAN, JOHN J., *Criminal Investigation.* New York: McGraw-Hill Book Co., 1974.

KIRK, PAUL L., *Crime Investigation.* New York: John Wiley & Sons, Inc., 1974.

—— and BRADFORD, LOWELL W., *The Crime Laboratory.* Springfield, Ill.: Charles C Thomas, Publisher, 1972.

KREMA, VACLAV, *The Identification and Registration of Firearms.* Springfield, Ill.: Charles C Thomas, Publisher, 1971.

MATHEWS, J. HOWARD, *Firearms Identification.* Springfield, Ill.: Charles C Thomas, Publisher, 1973.

MAURER, DAVID W. and VOGEL, VICTOR H., *Narcotics and Narcotic Addiction.* Springfield, Ill.: Charles C Thomas, Publisher, 1973.

MILLARD, JOHN T., *A Handbook on the Primary Identification of Revolvers and Semiautomatic Pistols.* Springfield, Ill.: Charles C Thomas, Publisher, 1974.

O'BRIEN, KEVIN P. and SULLIVAN, ROBERT C., *Criminalistics Theory and Practice.* Boston: Holbrook Press, Inc., 1972.

O'HARA, CHARLES E., *Fundamentals of Criminal Investigation.* Springfield, Ill.: Charles C Thomas, Publisher, 1974.

SANSONE, SAM J., *Modern Photography for Police and Firemen.* Cincinnati: The W. H. Anderson Co., 1971.

SVENSON, ARNE and WENDELL, OTTO, *Techniques of Crime Scene Investigation.* New York: American Elsevier Publishing Co., Inc., 1965.

WESTERN, PAUL B. and WELLS, KENNETH M., *Criminal Investigation,* 2nd Edition. Englewood Cliffs, N.J.: Prentice-Hall, Inc., 1974.

Index

A

Alcohol, 215
Amphetamines, 216–17

B

Barbiturates, 213–14
Bloodstain evidence:
    bleeding, 63
    color and visibility, 63
    samples, 67
    shade, persistency, and age, 63
Body fluids:
    identification, 69
    importance, 68–69
Bufotenine, 219
Bullet holes:
    angle, 177
    direction, 175
    distance, 177
    sequence, 176
    type of ammunition, 177

C

Caliber, 148
Cameras:
    fingerprint, 87
    motion picture, 87
    Speed Graphic, 86
    35 mm, 87
Camera traps, 101–2
Casting:
    definitions, 105
    materials, 105
    photographing, 106
    preparing, 108

Casting (cont.)
  preservation, 106
  recording, 106
  removing, 113
  sketching, 106
  in snow, 112
  under water, 112
Chain of custody, 18–20
Charred documents, 203
Chemical processing, 139
Cocaine, 215
Codeine, 212
Color photography, 96
Combination mark, 48
Comparison samples, 79
Crime scene photographing:
  arson, 99
  hangings, 100
  homicides, 99
  riots and disorders, 101
  surveillance, 101
  vehicle accidents, 100

D

Dangerous drugs, 206
Darvon, 214
Demerol, 213
Developer, 91
Diethyltryptamine (DET), 220
Dilaudid, 212
Dimethyltryptamine (DMT), 219
D-Lysergic Acid Diethylamide (LSD), 221–23
DOET, 220
DOM, 220
Drug abuse:
  abuser, 207
  addict, 208
  casual supplier, 208
  experimenter, 208
  supplier, 208
  user, 208
Dry-cleaning marks, 61
Dust, 78

E

Ejector, 149

Evidence:
  collecting, 78, 168
  custodian, 18
  definition, 16
  depositories, 22
  evaluating, 7
  fixed, 16
  fragile, 16
  laboratory examination request, 26
  ledger, 22
  movable, 16
  packaging, 80
  packing and wrapping, 24
  preserving, 7, 132, 169
  protection, 51
  receipt, 19
  releasing, 7–8
  room, 22
  subvoucher file, 22
  tagging, 7, 18
  transmitting to laboratory, 25
  voucher, 21
  voucher file, 22
Exposure meters, 89
Extortion, 190
Extractor, 148

F

Fibers:
  examination, 73
  transmittal to laboratory, 73–74
Film:
  contrast, 91
  development, 91
  orthochromatic, 89–90
  panchromatic, 90
  speed, 90
Filters, 92
Fingernail scrapings, 74
Fingerprint patterns:
  arches, 141
  loops, 142
  whorls, 143
Fingerprint records, 146
Fingerprints:
  characteristics, 120
  definitions, 121
  plain impressions, 125

recording, 123
records and forms, 121–22
rolled impressions, 124
Fluorescence of glass, 183–84
Footprints, 116
Forgery, 189
Friction mark, 48

G

Gas-Liquid chromatography, 84–85
Glass:
    collection, 168
    evaluation, 170
    evidence, 166–67
    fragments, 183
    general examination, 171

H

Hair:
    examination, 71–72
    importance, 70
Handrwritten standards, 198
Heat fractures, 179
Heroin, 209–10

I

Infrared light, 82
Infrared photography, 93–94
Inks, 203
Inventories and inspections, 22
Investigative notes, 12–14
Iodine method, 136–37

J

Jewelers' marks, 61

K

Key control, 22
Khat, 215

L

Latent fingerprints, 129–30
Latex rubber, 115
Laundry marks, 61
Legal photography, 102–3
Lens speed, 87
Liquid sulphur, 115

M

Marihuana, 206, 223–25
Measurements:
    indoor, 12
    outdoor, 12
Mescaline, 218–19
Methadone, 213
Methamphetamine, 217–18
Methylene Dioxyamphetamine (MDA), 220
Molding, 117
Morphine, 209
Moulage-Agar compositions, 115
Mutilated documents, 203

N

Negative impressions, 48
Neutron Activation Analysis (NAA), 164

O

Opium, 209

P

Paint spots, 182
Palm prints, 126
Papaverine, 213
Paregoric, 212
Peyote, 218–19
Phencyclidine (PCP), 220
Pistol, 148
Plaster and building materials, 77
Porous articles, 67

Powdering, 133–35
Proximity tests, 160
Psilocybin, 219
Psilocyn, 219

R

Refractive index, 184

S

Safe insulation, 77
Safety glass, 173
Seminal fluid, 69
Serial numbers, 59
　restoration, 160
Sexual assaults, 74
Sketches:
　finished, 9–10
　rough, 9
Shotgun, 148
Silicone rubber, 114
Soils, rocks, and minerals:
　difference, 76
　evidence, 76–77

Spectrograph, 83
Spectrophotometers, 84

T

Test firing, 162
Tire impressions, 117
Tone separation, 89
Toolmarks:
　casting, 57
　definition, 48
　evidence, 50
　identification, 50
　photographing, 57
　removal of, 55
Typewritten documents, 198
Typewritten exemplars, 199–200
Typewritten standards, 202

U

Ultraviolet light, 81
Ultraviolet photography, 93–96